Algae

Edited by Yee Keung Wong

Published in London, United Kingdom

IntechOpen

Supporting open minds since 2005

Algae
http://dx.doi.org/10.5772/intechopen.73417
Edited by Yee Keung Wong

Contributors
Maximilian Lackner, Erich Markl, Hannes Grünbichler, Leda Giannuzzi, Koji Iwamoto, Ayumi Minoda, Vetrivel Anguselvi, R E Masto, Ashis Mukherjee, P K Singh

Notice
Statements and opinions expressed in the chapters are these of the individual contributors and not necessarily those of the editors or publisher. No responsibility is accepted for the accuracy of information contained in the published chapters. The publisher assumes no responsibility for any damage or injury to persons or property arising out of the use of any materials, instructions, methods or ideas contained in the book.

First published in London, United Kingdom, 2019 by IntechOpen
IntechOpen is the global imprint of INTECHOPEN LIMITED, registered in England and Wales, registration number: 11086078, The Shard, 25th floor, 32 London Bridge Street
London, SE19SG – United Kingdom
Printed in Croatia

British Library Cataloguing-in-Publication Data
A catalogue record for this book is available from the British Library

Additional hard and PDF copies can be obtained from orders@intechopen.com

Algae
Edited by Yee Keung Wong
p. cm.
Print ISBN 978-1-83880-562-3
Online ISBN 978-1-83880-563-0
eBook (PDF) ISBN 978-1-83880-723-8

We are IntechOpen,
the world's leading publisher of
Open Access books
Built by scientists, for scientists

4,200+

Open access books available

116,000+

International authors and editors

125M+

Downloads

Our authors are among the

151

Countries delivered to

Top 1%

most cited scientists

12.2%

Contributors from top 500 universities

Interested in publishing with us?
Contact book.department@intechopen.com

Numbers displayed above are based on latest data collected.
For more information visit www.intechopen.com

Meet the editor

Dr. Wong Yee Keung is currently Assistant Professor in the Applied Science and Environmental Studies Team, School of Science & Technology, The Open University of Hong Kong. Moreover, he is Director of the "Centre of Excellence in Water Quality & Algal Research". Besides, Dr. Wong is a Fellow of Institute of Biomedical Science (FIBMS), member of the Institution of Environmental Sciences (IEnvSc), Royal Society of Biology (RSB), International Water Association (IWA), Association of Environmental Engineering and Science Professors (AEESP) and Hong Kong Biotechnology Organization (HKBIO). He is also Editorial Board Member of Current Research in Hydrology and Water Resources (Gavin Publishers) and Advances in Bioscience & Bioengineering (SciencePG). He published over 90 international conference papers, peer-reviewed journals, monographs, etc.

Contents

Preface

This Edited Volume is a collection of reviewed and relevant research chapters, concerning the developments within the "Algae" field of study. The book includes scholarly contributions by various authors and edited by a group of experts pertinent to Agricultural and Biological Sciences. Each contribution comes as a separate chapter complete in itself but directly related to the book's topics and objectives.

The book is divided in 4 chapters dealing with the following topics: Bioremediation of Biophilic Radionuclides by Algae, Cyanobacteria Growth Kinetics, Cyanobacteria for PHB Bioplastics Production: A Review and CO_2 Capture for Industries by Algae.

The target audience scientists, university and college professors, research professionals, students and users of academic libraries.

IntechOpen

Chapter 1

Bioremediation of Biophilic Radionuclides by Algae

Koji Iwamoto and Ayumi Minoda

Abstract

High amounts of radionuclides were released into the environment by the nuclear power plant accident of 2011 in Japan. Among the radioactive material, cesium, iodine, and strontium were especially dangerous because of their biophilic characteristics that allowed them to accumulate in living organisms, either as essential elements for iodine or analogs of potassium and calcium for cesium and strontium, respectively. As a result, there was a high social demand for decontamination to avoid exposure to these elements. The authors screened around 200 strains of algae and plants for their ability to absorb radioactive nuclides. The eustigmatophycean algae *Vacuoliviride crystalliferum* and the cyanophytes *Stigonema ocellatum* and *Nostoc commune* showed the highest bioaccumulation activity for the removal of cesium, strontium, and iodine from the environment, respectively. In addition to these strains, the authors also found that the extremophilic unicellular red algae *Galdieria sulphuraria* could remove high levels of dissolved cesium from media in mixotrophic growth conditions. In this chapter, the intake mechanism of cesium, iodine, and strontium is reviewed. Recent findings on the absorption of these elements by algae are discussed to highlight the possibility of decontaminating polluted land and water at nuclear sites by phytoremediation.

Keywords: radionuclides, phytoremediation, cesium, strontium, iodine, *Vacuoliviride crystalliferum*, *Galdieria sulphuraria*

1. Introduction

On March 11, 2011, Japan suffered a large earthquake, known as the Great East Japan Earthquake, causing a tsunami that damaged the Fukushima 1st Nuclear Power Plant (F1NPP). This damage led to release of a large amount of radioactive matter into the environment, estimated at 11.6 EBq (exa Bq: 10^{18} Bq) [1]. Radioactive xenon (Xe-133) was the most abundant material discharged, at 11.3 EBq, followed by iodine (I-131, I-132, I-133, and I-135), tellurium (Te-127m, Te-129m, Te-131m, and Te-132), and cesium (Cs-134 and Cs-137). In addition to these, radioactive strontium (Sr-89 and Sr-90), barium (Ba-140), yttrium (Y-91), and plutonium (Pu-238, Pu-239, Pu-240, Pu-241) were also released among others. Among these radionuclides, cesium, strontium, and iodine gained the most attention because of their biophilic properties, in addition to the high amount of the discharged activity, which was 33 PBq (peta Bq: 10^{15} Bq), 200 PBq, and 2.1 PBq in total, 10–37 PBq, 150 TBq (tera Bq: 10^{12} Bq), and 90–500 PBq in the air, and 1.9 PBq, 90–500 TBq, and 2.7 PBq in the ocean, respectively [1–5]. However, eventually the attention focused solely on cesium because iodine and Sr-98 have relatively short half-lives and so their levels decreased to below detectible levels at an early stage (50.5 days for Sr-98 and 8 days for I-131, which is the longest half-life period among radioactive iodine released by the accident) [1]. For strontium, in addition to the early decrease in the contamination levels of Sr-98, the total discharged activity of Sr-90 was comparatively small, and the polluted area was limited to near the sur-roundings of the F1NPP due to its low volatility [6]. However, radioactive strontium and iodine still need to be monitored and decontaminated because 140 TBq of Sr-90, with a half-life 29.1 years, was released into environment, and I-129, with a half-life of 16 million years, could be produced from 3.3 PBq of Te-129m in the polluted environment [7, 8]. I-129 is derived from Te-129m, and small amounts of I-129 produced in the reactor were also released during the accident [9].

High-level radiation-contaminated water in the reactor building and the acces-sory facilities of F1NPP was treated by various chemical and physiological methods including the SARRY system (simplified active water retrieve and recovery system) for cesium removal and the ASPS system (advanced liquid processing system) for multinuclide removal [12, 13]. However, the decontamination of the surround-ing area was not as simple because of the wide area of contamination and the low concentrations of the radioactive materials.

However, the total mass of the radioactive material was not as high as the total activity from F1NPP, which was calculated by following equation (**Table 1**) [10, 11]:

$$W \ = \ B \times 8.62 \times 10^{21} \times M \times T \tag{1}$$

where w is the total mass (g), B is the total activity (Bq), M is the radionuclide atomic weight, and T is the half-life period (hour).

Even when calculating the total mass of all the biophilic elements, it did not exceed 6 kg. On the other hand, the polluted area exceeded over 10,000 km^2, which included forests, field, lakes, rivers, and houses, with a rate of 0.5 $\mu Sv \ h^{-1}$ in the air recorded on April 1, 2011 [14]. Ten years later, the area would still cover around 3000 km^2 due to the half-life period of Cs-137, which is 30 years [14]. The physiological and chemical methods for decontamination are costly and difficult to apply for such a weak and widespread contamination. Hence, the biological method, called bioremediation, is a good candidate for decontamination [10]. Bioremediation is a method that uses living organisms to accumulate or degrade a contaminating material. This method has many advantages because it allows for the possibility to treat various pollutants thanks to its biological diversity. It is effective

Nuclide	Half-life period	Total activity (Bq)	Mass (g)*
Cs-134	2.1 years	1.8×10^{16}	373
Cs-137	30.0 years	1.5×10^{16}	4753
Sr-89	50.5 days	2.0×10^{15}	1.8
Sr-90	29.1 years	1.4×10^{14}	27.7
I-131	8.0 days	1.6×10^{17}	35.5
I-132	2.3 hours	1.3×10^{13}	3.4×10^{-5}
I-133	20.8 hours	4.2×10^{16}	1.0
I-135	6.6 hours	2.3×10^{15}	1.7×10^{-2}

Mass was calculated from the total activity, radionuclide atomic weight, and half-life using the equation in the text below.

Table 1.
Total activity and the mass of radioactive cesium, strontium, and iodine released from F1NPP.

for trace levels of pollutants; it can reduce the cost of remediation because of the low energy usage. However, in addition to the disadvantage of it being both temperature- and weather-dependent, the speed of the remediation method is limited. Despite this, bioremediation is considered to be the remediation technology for the next generation.

2. Uptake of cesium, strontium, and iodine by organisms

Cesium, strontium, and iodine were regarded as the most hazardous of the radionuclides diffused into the environment by the F1NPP accident. These elements are easily absorbed by living organisms through water and air or indirectly through food and result in an increased health risk by internal exposure. The indirect uptake of these nuclides through food is more serious because the strength of the activity may be increased by biological concentration via food chain [10]. Therefore, the number of radionuclides, such as Cs-137, Sr-90, and I-131, as well as Ru-106 and K-40, has been monitored by the United States (Food and Drug Administration), Japan (Ministry of Health, Labor and Welfare), and other countries in addition to toxic metals, pesticides, and chemicals [11].

2.1 Cesium

Of the different radionuclides, radioactive cesium has attracted the most attention for the following reasons: (i) the total amount of released radioactivity for Cs-134 and Cs-137 was very high, (ii) the long half-life of cesium (2.1 and 30.0 years for Cs-134 and Cs-137, respectively) means it remains in environment for many years, (iii) the area contaminated by cesium was considerably large with respect to its volatility, and (iv) cesium is easily absorbed and accumulated in the body of living organisms.

Cesium is absorbed by the body as an analog of potassium, which is one of the essential elements of living cells, and is accumulated in the parenchyma of the muscles and organs [12]. However, the biological concentration factor of cesium is not very high, unlike for heavy metals, and is in the range of one to two orders of magnitude in large predatory fish [13–19]. The effective half-life is considered to be the same level as biological half-life in humans, about 100 days, because of the long half-life of Cs-134 and Cs-137 [20]. Cesium is thought to be taken up by the potassium assimilation system, which includes the potassium transporter, the sodium/potassium pump, and the potassium channel. In animals, cesium is absorbed in the intestinum tenue, stored partially in the liver, and used in the cells and organs across the whole body, including muscles, bones, brain, etc. In plants, potassium also has important physiological roles, such as the extension and division of the cell, the opening and closing of stoma, and signal transduction between organelles. On the other hand, it has been previously reported that over 200 μM of cesium can inhibit plant growth by disturbing the potassium uptake and jasmonate signaling [21, 22]. Many potassium transporting systems have been identified, and cation and/or calcium-transporting systems are thought to be involved in cesium transport in plants [23].

Figure 1 is a schematic diagram of a charging and discharging system of cesium in a plant root cell [23, 24]. Plant roots have several reported systems for potassium intake, including K^+/H^+ symporters (KUP) and inward-rectifying K^+ channels (KIRCs) [25]. Cesium is transported into the cell by these systems. The KUP system belongs to the KT/KUP/HAK family. AtHAK5 and AtKUP/HAK/KT9 were identified in *Arabidopsis thaliana* [26]. The KUP system is thought to be involved in the

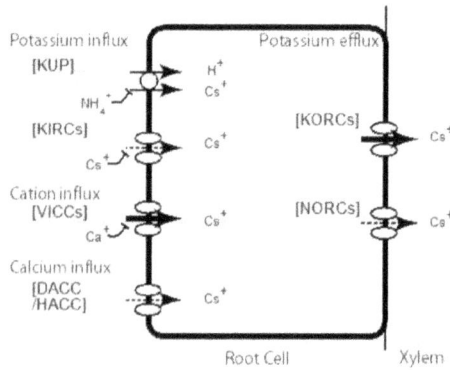

Figure 1.
Cesium charging and discharging system in a plant root cell. DACCs: depolarization-activated Ca^{2+} channel, HACC: hyperpolarization-activated Ca^{2+} channel, KIRC: inward-rectifying K^+ channels, KORC: outward-rectifying K^+ channels, KUP: K^+ uptake, NORC: nonselective outward-rectifying K^+ channel, VICC: voltage-independent cation channel. The arrows and their widths indicate the direction of transport and their relative activity, respectively.

uptake of cesium when potassium levels are low because the gene expression of AtHAK5 has been reported to increase in potassium or NH_4^+-deficient conditions. Moreover, KUP has been previously shown to transport cesium by an AtHAK5-deficient *Arabidopsis* mutant and an AtHAK5-producing yeast whose phenotypes showed lower Cs+ accumulation and higher cesium uptake, respectively [26]. The role of KIRCs in cesium uptake is not clear because these potassium channels are severely inhibited by the transport of Cs ions, which act as potassium channel blockers and close the channel [23, 27].

The importation of Cs in plants is also carried out by other cation efflux systems. One of these are the voltage-independent cation channels (VICCs) [23]. The AtCNGC (*A. thaliana* cyclic-nucleotide gated channel) and AtGLR (*A. thaliana* glutamate receptor) gene families are involved in the VICCs identified in *A. thaliana* [24]. Other cation efflux systems also considered to be involved in cesium transport are HACCs (hyperpolarization-activated Ca^{2+} channel for calcium efflux), DACCs (depolarization-activated Ca^{2+} channel systems for calcium efflux), KCO (the cation-transporting channel to vacuole), KEAs (K^+ efflux antiporter as potassium-proton antiporters), and AtNHX1 (*A. thaliana* Na^+/H^+ antiporter classified in cation-proton antiporters) [23]. For discharging of cesium from cells to bundle, KORCs (outward-rectifying K channel) are believed to play a fundamental role among the potassium efflux system such as NORCs (nonselective outward-rectifying K channel) [23].

2.2 Strontium

Radioactive strontium received the least attention of the three biophilic elements released by the F1NPP because the total discharged radioactivity was the lowest, at 2.1 PBq, that is, 1/10th of that of radioactive cesium. Moreover, Sr-89, with a half-life of 50.5 days, made up about 2 PBq of this, and hence the majority of the strontium in the environment decayed early on. The area polluted by radioactive strontium was limited because of its nonvolatility, and detection was difficult because both Sr-89 and Sr-90 have low-range radiation as low-energy beta-ray nuclides [1, 10]. Strontium is taken up by cells in the intestine as an analog of calcium and is stored in the bones [28]. Once strontium is absorbed, it is not easily excreted from the body due to its long biological half-life of 30–50 years. This poses a serious health risk due to long-term internal exposure to radiation. Furthermore, high levels of strontium were reported in predatory freshwater fish, with a concentration factor of 10^3 [29]. Therefore, radioactive strontium should receive equal if not greater attention than the other radionuclides. Calcium is the most abundant metal element and mostly exists in bones in animals. In plants, calcium is the second most abundant metal element after potassium and exists in the extracellular space in its insoluble form. It binds to cell wall components to form part of the support system of the plant structure. In the cell, calcium is stored in organelles such as mitochondria and endoplasmic reticulum, or chloroplasts and vacuoles in plants. The concentration of calcium in the cytosol is kept in at very low levels (10^{-7} M) because calcium plays an important role for signal transduction in cells and acts as a secondary messenger. Therefore, there are constitutive calcium excretion systems on the surface of the cell, including CAX1 and CAX2 as the H^+/Ca^{2+} antiporter, ECA and ACA as the calcium pump, and TPC1 and VICCs as the ion channels [30].

Figure 2 is a schematic diagram of the absorption mechanism of calcium in the small intestine of animals. There are two types of absorption, paracellular absorption and transcellular absorption, which function to transport calcium passively under high calcium concentrations and actively under low calcium concentrations, respectively [31]. In the paracellular pathway, small molecules and ions such as Ca^{2+} are selectively allowed to a pass through a very narrow space between the cells of the intestinal epithelium, i.e., the tight junctions [32]. For transcellular absorption, there are two active pathways for Ca^{2+} absorption. One is mediated by a calcium channel called Ca_V 1.3, and the other is mediated by a calcium transporter called TRPV6. The calcium transported into the intestinal epithelium cell by Ca_V 1.3 is thought to be released into the bloodstream vesicles. On the other hand, calcium incorporated into the cell is secreted into the plasma by the vesicular transport pathway or via PMCA1b, a calcium pump (Ca^{2+} ATPase). Calcium ions are delivered to the pump by simple diffusion of free Ca^{2+}

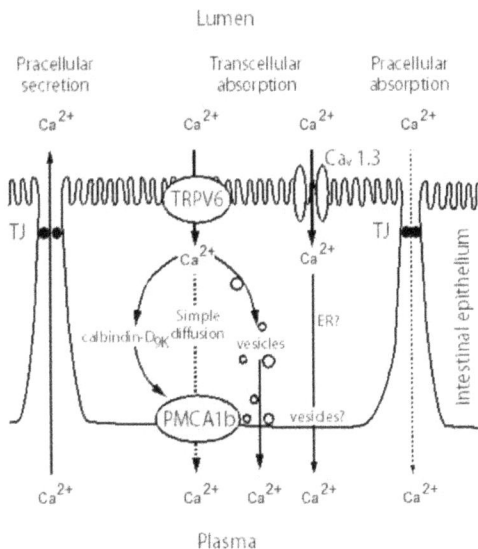

Figure 2.
The calcium uptake system in the small intestine of animals. The arrows and their widths indicate the direction of transport and their relative activity, respectively. ER: endoplasmic reticulum, TJ: tight junction. Solid and dashed lines indicate active transport and diffusion, respectively.

or by the facilitated diffusion pathway with a vitamin D-dependent calcium-binding protein, calbindin-D9K [31].

2.3 Iodine

Radioactive iodine is easily absorbed by the body as the main component of the thyroid hormone. Iodine is focally accumulated and concentrated at the thyroid gland, which increases the risk of developing goiter, especially in infants and small children. As a result of the F1NPP accident, radioiodide was spread into a wide area due to its high volatility. Radioactivity was detected not only in many places and food but also in the drinking water, because of its high solubility. For these reasons, it attracted the most attention immediately after the F1NPP accident. After several months, the attention shifted to radioactive cesium due to the early decay of iodine as a result of its short half-life (half-life of 8.0 days for I-133). However, radioactive iodine I-129 still poses a great health risk as it was both directly released from the reactor and can be formed indirectly via Te-129m [7–9]. Even though the release of I-129 was small and the concentration is currently not at a level that would pose a risk to human health [33, 34], because of its extremely long half-life of 16 M years, it could be deposited in the long-term by its continuous release from the nuclear power station or nuclear fuel reprocessing facilities, by another serious accident in the nuclear facility, or by the use of the nuclear weapons.

In higher animals, iodine is concentrated and located in the thyroid glands. Iodine is incorporated into the thyroid glands during the biosynthesis of thyroid hormone, namely triiodothyronine (T3) and thyroxine (T4). Iodide ions (I^-) brought to thyroid epithelial cells by the blood are oxidized into iodinium ions (I^+) by thyroperoxidase. This is activated by the thyroid-stimulating hormone (TSH). The subsequent iodination of tyrosyl residues of the thyroglobulin protein produces iodotyrosines, monoiodotyrosine (MIT), and diiodotyrosine (DIT) by an electrophilic substitution reaction. T4 and T3 are then synthesized and stored in thyroglobulin following an MIT and DIT coupling reaction. Finally, the thyroid hormone-binding protein is hydrolyzed into water-soluble amino acids by the activation of TSH [35].

Brown algae are known to accumulate iodine at rates 10,000 times more than that in seawater. **Figure 3** shows the schematic diagram of the absorption mechanism of iodine by brown algae [36]. Iodide ions (I^-) in the seawater are oxidized

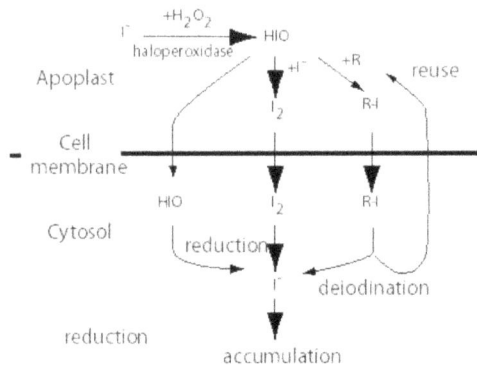

Figure 3.
Iodine uptake system in the brown algae.

into hypoiodous acid (HIO) by haloperoxidase in the apoplastic space in the cell wall. Oxidized iodine passes through the cell membrane, by facilitated diffusion in HIO form, or reacts with other I^- or organic substances to produce molecular I_2 or organic iodine, respectively, and crosses the membrane by diffusion. The iodine transported as HIO, I_2, or organic iodine is then reduced or deiodinated in the cytosol and stored in the cell in I^- form, binding noncovalently with carbohydrates, polyphenols, and proteins [33].

3. Phytoremediation by higher plants for decontamination of a radioactive element

Bioremediation using plants is called phytoremediation, and its use for the decontamination of radioactive elements has been previously reported. For cesium decontamination, Dushenkov et al. reported on the bioaccumulation of cesium, strontium, and uranium by sunflowers in which over 90% of the polluting substances were removed from a solution within 24, 28 and 24 hours, respectively [37]. Broadley et al. showed that dicotyledons (*Magnoliopsida*) possessed a higher activity of bioaccumulation than monocotyledons (*Liliopsida*). Specifically, from the comparison of data extracted from 14 reports regarding cesium bioaccumulation in the plant shoots of 136 species among Magnoliophyta taxa, using a residual maximum likelihood (REML) analysis, they found that a redroot pigweed (*Amaranthus retroflexus*) in Amaranthaceae had the highest bioaccumulation ability [38]. The authors then found that a sugar beet (*Beta vulgaris*) in Chenopodiaceae, a turnip (*Brassica napus*) in Brassicaceae, quinoa (*Chenopodium quinoa*) in Chenopodiaceae, and a swede (*Brassica napus*) had the highest abilities for cesium accumulation after *A. retroflexus*.

The bioaccumulation of strontium by higher plants was surveyed comprehensively among 670 species in 138 families [39], although the individual species names were not specified. Sasmaz and Sasmaz selected *Euphorbia macroclada* in Euphorbiaceae, *Verbascum cheiranthifolium* in Scrophulariaceae, and *Astragalus gummifer* in Fabaceae from a polluted field in a mining area in Eastern Turkey for potential use in phytoremediation [40]. These reports indicate that these plants have a great potential for the decontamination of radiostrontium from polluted areas. Specifically, redroot pigweed and a tepary bean (*Phaseolus acutifolius*) have shown rates of 4.5% and 3.1% removal of Sr-90 from soil, respectively [41].

There are few reports on the uptake of iodine by higher plants because iodine is not an essential element for growth. Instead, it is considered to be a toxic element that causes growth inhibition. For example, type III Akagare disease is caused by the inhibition of photosynthesis due to high concentrations of iodide (>1 ppm) [42]. Therefore, the screening of iodine tolerant plants may be required for the effective phytoremediation and subsequent decontamination of radioactive iodine, unlike cesium and strontium.

Subsequently, decontamination was tested using higher plants, including sunflowers and 12 plant species in Fabaceae and Poaceae, to remove radioactive cesium from contaminated soils in the area polluted by the F1NPP accident [43]. However, a significant rate of removal was not achieved because only 1/2000 of the radioactive cesium in the soil was removed by sunflower [44]. Phytoremediation by higher plants is considered to be an effective method for the decontamination of low- and middle-level radio-polluted soils. However, the screening and breeding of a radionuclide hyperaccumulator is required, in addition to the optimization of the culture conditions and application, according to the localization of the target nuclides [43].

4. Phytoremediation of biophilic radionuclides by algae

4.1 Advantages of microalgae for decontamination of water

Phytoremediation using algae, also called phycoremediation, has the following advantages, in particular when using microalgae [45]:

1. Fast remediation because of the high growth rates of microalga.

2. Remediation with lower energy costs because of autotrophy.

3. High and effective remediation because of the biological concentration function.

4. Volume reduction of polluted material/water because of the single cellular or simple structure.

In particular, volume reduction by microalgae is considered to be an important advantage in the F1NPP case.

4.2 Algae absorb radioactive cesium, strontium, and iodine

The ability to remove the radioactive cesium, strontium, and iodine from the solution was comprehensively examined to cover broad phylogenic variation by Fukuda et al. [46]. These aquatic plants and algae were consisted from 188 strains of algae and aquatic plants including 91 seawater, 86 freshwater, and 11 terrestrial strains, which covered almost all phylogenic group, i.e., 45 classes, 21 divisions, and 7 super groups such as cyanobacteria, Opisthokonta, Excavata, Archaeplastida, Rhizaria, Alveolata, and Stramenopiles in two domains (**Table 2**).

These organisms show various advantageous features in their morphology, physiology, biochemical properties, and nutritional properties, namely autotrophy and/or heterotrophy. Most of these strains were obtained from the culture collections of the Laboratory of Plant Diversity and Evolutionary Cell Biology, University of Tsukuba, Japan, and the Microbial Culture Collection at the National Institute for Environmental Studies (NIES Collection, Tsukuba, Japan). Several other strains were collected from the surrounding area of the University of Tsukuba or were purchased from the local market.

The strains were inoculated in a 70 mL scale plastic culture bottle containing 15 mL of medium under a fluorescent light at 20°C. The strains were inoculated in 15 mL of fresh medium in the plastic culture bottle 1 day before the addition of artificial radionuclide, Cs-137, Sr-85, or I-125, with a concentration of 1 kBq/mL. The same medium was used for the preparative and the experimental culture, except for the cesium removal assay in which potassium was excluded. An aliquot was obtained, and the radioactivity contained in the medium fraction and the cell fraction was assayed using a gamma-ray counter by silicone oil layer methods [47]. The medium fraction was only assayed for the macroscopic algae and aquatic plants, which do not fit the silicone oil layer methods.

For the comprehensive examination of the radionuclide elimination ability of algae and aquatic plants, 167 strains out of 188 strains showed this activity (**Figure 4a**). From the 167 strains, the 15 strains that showed an elimination activity of over 40% were reexamined in a second screening (**Figure 4b**). As a result of the second screening, four strains possessing an elimination activity of 30% were selected as high radioactive cesium eliminators. These were the freshwater eustigmatophycean algae *Vacuoliviride*

Domain	Super group	Divisions	Nutrient condition
Prokarya	Bacteria	Cyanobacteria	A
Eukarya	Opisthokonta	Fungi	A/H
		Choanozoa	H
	Excavata	Metamonada	H
		Euglenozoa	A/H
		Percolozoa	H
	Archaeplastida	Rhodophyta	A
		Chlorophyta	A
		Mesostigmatophyta	A
		Chlorokybophyta	A
		Klebsormidiophyta	A
		Zygnematophyta	A
		Charophyta	A
		Land plants	A
		Cryptophyta	A/H
	Hacrobia	Haptophyta	A
	Rhizaria	Cercozoa	A/H
	Alveolata	Dinophyta	A/H
	Stramenopiles	Bicosoecacea	H
		Pseudofungi	H
		Ochrophyta	A/H

A, H, and A/H, nutrient conditions indicating "mostly autotrophic," "mostly heterotrophic," and "a mixture of A and H," respectively.

Table 2.
The phylogenic position of experimental organisms (phylum level).

crystalliferum (strain nak 9, 90% elimination), the freshwater florideophycean algae *Batrachospermum virgato-decaisneanum* (NIES-1458, ca. 38% elimination), and two strains of aquatic plants (tracheophytes) *Lemna aoukikusa* (TIR 2 and TIR 3, ca. 45 and 66% elimination, respectively). Of these five strains, *V. crystalliferum* showed the highest elimination ability, removing over 90% of the cesium-137 from the liquid within 2 days. This alga is considered to be a candidate strain for the decontamination of radiocesium by phytoremediation. The floating weed *L. aoukikusa* is also considered to be useful for the decontamination of water because it is easy to harvest using a net.

For the elimination of strontium, activity was confirmed in 181 of 188 strains (**Figure 5a**). By the second screening, 10 strains were identified, which had an elimination ability over 30% (**Figure 5b**). From these, three strains were selected as the highest strontium eliminators, namely the freshwater cyanobacterium *Stigonema ocellatum* (NIES-2131, ca. 41% elimination), the freshwater chlorophycean alga *Oedogonium* sp. (nak 1001, ca. 36% elimination), and the freshwater Magnoliopsida *Egeria densa* (We2, ca. 34% elimination).

For the elimination of radioactive iodine, all strains except the green alga *Stigeoclonium aestivale* (NIES-531) showed activity at the global screening. However, the level varied between 0% and 40% depending on the strain (**Figure 6a**). After the second screening, 14 strains that showed an iodine elimination activity of over 40% were found, and finally, four high radioiodine eliminator strains were selected

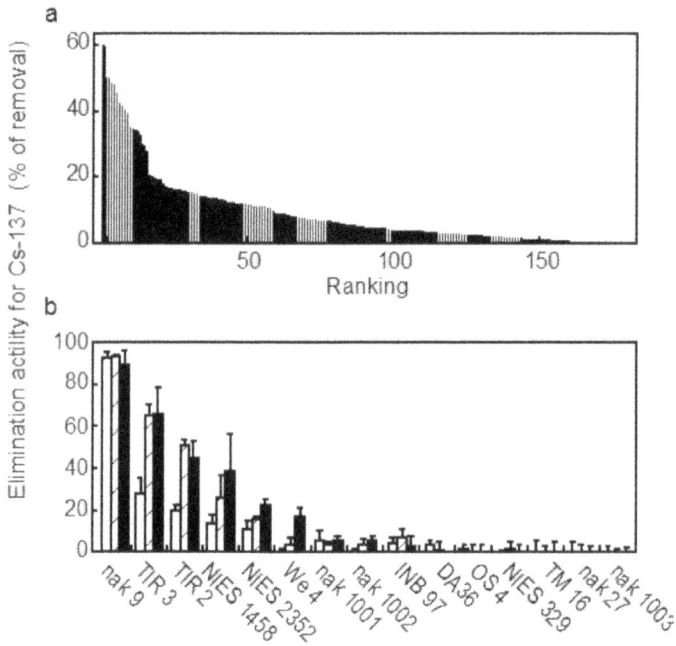

Figure 4.
Elimination activity of Cs-137 from the culture medium by algae and aquatic plants. (a) Global screening. Average values were ranked in descending order. (b) Second screening by selected algae and aquatic plants.

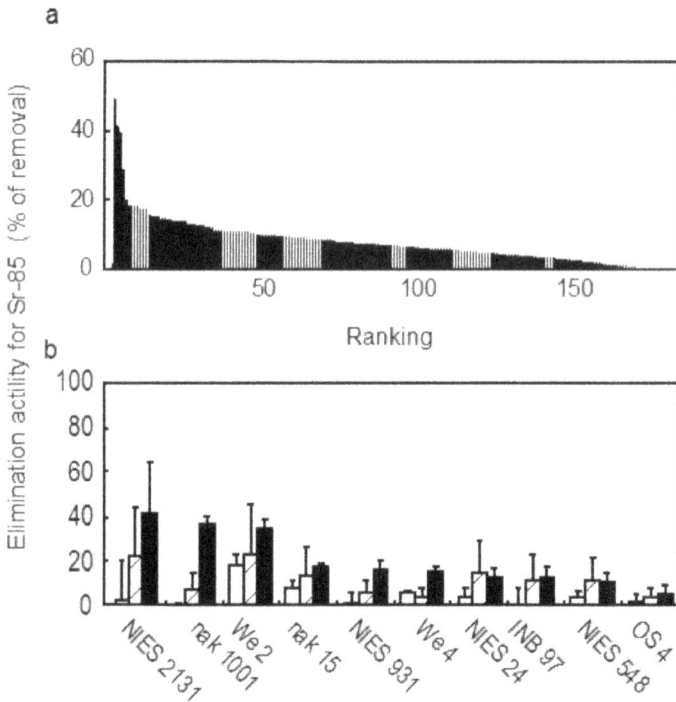

Figure 5.
The elimination activity of Sr-85 from the culture medium by algae and aquatic plants. (a) Global screening. Average values are ranked in descending order. (b) Second screening by selected algae and aquatic plants by the global screening.

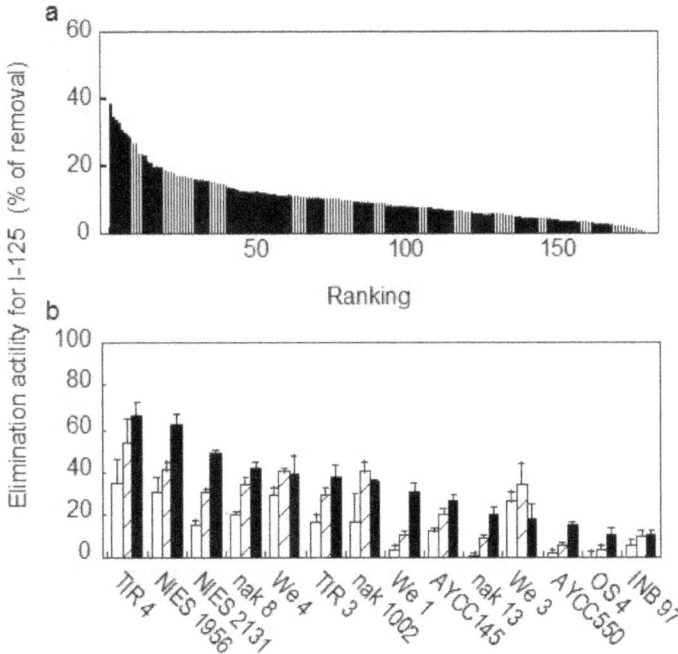

Figure 6.
Elimination activity of I-125 from the culture medium by algae and aquatic plants. (a) Global screening. Average values are ranked in descending order. (b) Second screening by selected algae and aquatic plants by the global screening.

(**Figure 6b**). These strains were the terrestrial cyanobacteria *Nostoc commune* (TIR 4, ca. 66% elimination) and *Scytonema javanicum* (NIES-1956, ca. 62% velimination), the freshwater cyanobacterium *Stigonema ocellatum* (NIES-2131, ca. 49% elimination), and the freshwater xanthophycean alga *Ophiocytium* sp. (nak 8, ca. 42% elimination).

All the selected strains were either freshwater or terrestrial strains. This is apt because of the competitive inhibition of the stable elements in the seawater medium, which contained 10, 10 mM, and 0.5 µM of potassium, calcium, and iodine, respectively [48, 49]. Likewise, for heterotrophic algae, the absorption and elimination of radionuclides were inhibited by the potassium and calcium in the yeast extract used for the medium (data not shown). Interestingly, the cyanobacterium *S. ocellatum* (NIES-2131) possessed a high eliminating activity for both Cs-137 and I-125. This may suggest that the decontamination of both elements occurs simultaneously. These data suggest that algae are a key organism for the phytoremediation of a polluted environment, especially for water polluted by radioactive cesium.

4.3 High-cesium bioaccumulating alga *V. crystalliferum* and *Galdieria sulphuraria*

The high-cesium bioaccumulating alga *V. crystalliferum* was isolated and recently identified as a new species and genus [50]. At the cellular level, the cells of this alga possessed a typical structure, which is a large reddish globule and crystalline. The uptake of Cs-137 was quite significant. The time course of Cs-137 elimination from the medium is shown in **Figure 7**. Over 60% of radioactive cesium was eliminated within 15 min and 90% within 1 hour. The maximum velocity for uptake was calculated as 63 mg Cs-137 mg^{-1} Chl h^{-1}. It has been reported that sunflowers and the vetiver *Vetiveria zizanioides* absorbed 150 µg and 61% of cesium after 100 and

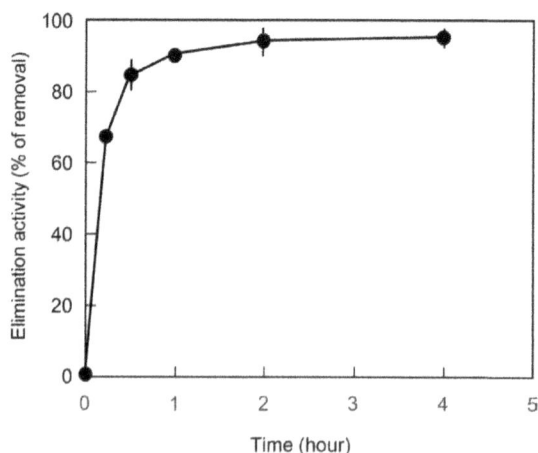

Figure 7.
Elimination of Cs-137 from the medium by V. crystalliferum.

168 hours, respectively [37, 51]. The partition coefficient (K_d) was calculated at around 4×10^5 L kg^{-1}. This value was higher than the value of cesium absorption by zeolite and Prussian blue, the ferrocyanide pigment [10]. Therefore, this alga could be used for the effective elimination and absorption of radioactive cesium. Another advantage was in the physiological toughness of the cell. It can be grown in severe environmental conditions, such as high osmotic pressure, dim light, and high and low temperatures, because it was originally isolated from sediment found in a bottle of glue [10, 50]. As such, the use of this organism could help to reduce the cost of phytoremediation for decontamination of radioactive cesium because the energy needed to maintain the culture conditions of this strain is minimal.

In addition to *V. crystalliferum*, Fukuda and his coresearchers found that an extremophilic unicellular red alga *Galdieria sulphuraria* eliminated ca. 50% of cesium in the medium in 10 days [52]. Interestingly, the elimination did not take

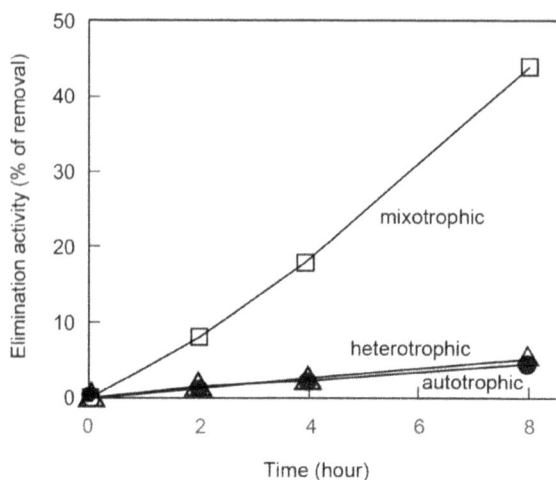

Figure 8.
Elimination of Cs-137 from the medium by G. sulphuraria. Cells were incubated without 25 mM glucose under light condition (autotrophic condition, closed circle), with 25 mM glucose under light condition (mixotrophic condition, open square), or with 25 mM glucose under dark condition (heterotrophic condition, open triangle).

place under autotrophic and heterotrophic conditions but only under mixotrophic conditions (**Figure 8**). Even though the elimination velocity was lower than that of *V. crystalliferum*, this alga may be useful for acidic contaminated polluted water because this alga prefers acidic conditions below pH 4.0.

5. Conclusion

In this chapter, we introduced a biological mechanism for the uptake of cesium, strontium, and iodine. Phytoremediation is a method that utilizes biological functions such as substance accumulation by biological concentration and biodegradation. Screening for a hyperaccumulator was carried out to apply phytoremediation technology in the decontamination of polluted land and water by radioactive substances released by the F1NPP accident. Global and secondary screening was used to identify several candidates of algae and aquatic plants for the phytoremediation and elimination of cesium, strontium, and iodine, respectively. In particular for cesium decontamination, the high-cesium bioaccumulating alga, novel eustigmatophytes *V. crystalliferum*, and an extremophilic unicellular red alga, *G. sulphuraria*, were found to possess high activity for elimination. Notably, the *V. crystalliferum* activity was particularly high. This suggests that these algae possess a specific transport system for cesium. One alternative could be to identify the transport system and introduce it into another plant or organism. In addition to the high activity, both algae are able to grow in severe environmental conditions, which is one of the required characteristics for phytoremediation. However, the decontamination of soil polluted by radioactive cesium would require more than only phytoremediation technology since the radioactive cesium found in and around the Fukushima area is firmly bound to the soil [53]. Therefore, further developments in the technology for the removal and solubilization of radioactive cesium in soils are essential. For this, the collaboration between scientists in different scientific disciplines is essential.

Author details

Koji Iwamoto* and Ayumi Minoda
Malaysia Japan International Institute of Technology, University Technology Malaysia, Kuala Lumpur, Malaysia

*Address all correspondence to: k.iwamoto@utm.my

IntechOpen

References

[1] METI (Ministry of Economy, Trade and Industry). Correction of Data on Radioactive Substance Emissions. News Release 20 October 2011 (in Japanese)

[2] Chino M, Nakayama H, Nagai H, et al. Preliminary estimation of release amount of ^{131}I and ^{137}Cs accidentally discharged from the Fukushima Daiichi nuclear power plant into the atmosphere. Journal of Nuclear Science and Technology. 2011;**48**:1129-1134

[3] Nuclear Emergency Response Headquarters. Report of Japanese Government to the IAEA Ministerial Conference on Nuclear Safety—The Accident at TEPCO's Fukushima Nuclear Power Stations. 2011. Available from: http://www.kantei.go.jp/foreign/kan/topics/201106/iaea_houkokusho_e.html [Accessed: May 26, 2018]

[4] Stohl A, Seibert P, Wotawa G, Arnold D, Burkhart JF, Eckhardt S, et al. Xenon-133 and caesium-137 release into the atmosphere from the Fukushima Dai-ichi nuclear power plant: Determination of the source term, atmospheric dispersion, and deposition. Atmospheric Chemistry and Physics. 2012;**12**:2313-2343

[5] TEPCO (The Tokyo Electric Power Co., Inc.). Estimation of the Total Amount of Radioactive Materials Released to the Atmosphere by the Accident of the Fukushima Daiichi Nuclear Power Plant. Press Release 24 May 2012 (in Japanese)

[6] Steinhauser G, Schauer V, Shozugawa K. Concentration of strontium-90 at selected hot spots in Japan. PLoS One. 2013;**8**:e57760

[7] Ramayya AV, Yoshizawa Y, Mitchell ACG. Level scheme of I^{129}. Nuclear Physics. 1964;**56**:129-139

[8] Taira Y, Hayashida N, Yamashita S, Kudo T, Matsuda N, Takahashi J, et al. Environmental contamination and external radiation dose rates from radionuclides released from the Fukushima nuclear power plant. Radiation Protection Dosimetry. 2012;**151**:537-545

[9] Miyake Y, Matsuzaki H, Fujiwara T, Saito T, Yamagata T, Honda M, et al. Isotopic ratio of radioactive iodine (^{129}I/^{131}I) released from Fukushima Daiichi NPP accident. Geochemical Journal. 2012;**46**:327-333

[10] Iwamoto K, Shiraiwa Y. Cesium accumulation by aquatic plants and algae. In: Gupta DK, Walther C, editors. Impact of Cesium on Plants and the Environment. Cham, Switerland: Springer International Publishing; 2016. pp. 171-185. DOI: 10.1007/978-3-319-41525-3_10

[11] Groch MW. Radioactive decay. Radio Graphics. 1998;**18**:1247-1256

[12] Tusa E. Cesium and strontium removal with highly selective ion exchange media in Fukushima and cesium removal with less selective media-14018. In: WM2014 Conference. Phoenix: Phoenix Convention Center; 2014

[13] TEPCO (The Tokyo Electric Power Co., Inc.). Situation of Storing and Treatment of Accumulated Water Including Highly Concentrated Radioactive Materials at Fukushima Daiichi Nuclear Power Station (354th Release). Press Release 28 May 2018

[14] Saito MU, Doko T, Koike F. Forecasting radiation effects on wildlife in Japan after the Fukushima nuclear accident, based on limited information of post-accident early stage in 2011. The International Archives of

the Photogrammetry, Remote Sensing and Spatial Information Sciences. 2014;**XL-2**:13-20. DOI: 10.5194/isprsarchives-XL-2-13-2014

[15] Friday GP, Cummins CL, Schwartzman AL. Radiological Bioconcentration Factors for Aquatic, Terrestrial, and Wetland Ecosystems at the Savannah River Site. Ailen, South Carolina: Westinghouse Savannah River Company; 1996. DOI: 10.2172/573692

[16] Betsy A, Sudershan Rao V, Polasa K. Evolution of approaches in conducting total diet studies. Journal of Applied Toxicology. 2012;**32**:765-776

[17] Arai T. Radioactive cesium accumulation in freshwater fishes after the Fukushima nuclear accident. Springer Plus. 2014;**3**:479. DOI: 10.1186/2193-1801-3-479

[18] Mizuno T, Kubo H. Overview of active cesium contamination of freshwater fish in Fukushima and eastern Japan. Scientific Reports. 2013;**3**:1742. DOI: 10.1038/srep01742

[19] McCreedy CD, Jagoe CH, Glickman LT, Brisbin IL Jr. Bioaccumulation of cesium-137 in yellow bullhead catfish (*Ameiurus natalis*) inhabiting an abandoned nuclear reactor reservoir. Environmental Toxicology and Chemistry. 1997;**16**:328-335. DOI: 10.1002/etc.5620160231

[20] Uchiyama M. Re-evaluation of the biological half-time of caesium in Japanese male adults. Journal of Environmental Radioactivity. 1998;**41**:83-94

[21] Adams E, Chaban V, Khandelia H, Shin R. Selective chemical binding enhances cesium tolerance in plants through inhibition of cesium uptake. Scientific Reports. 2015;**5**:8842. DOI: 10.1038/srep08842

[22] Adams E, Abdollahi P, Shin R. Cesium inhibits plant growth through jasmonate signaling in *Arabidopsis thaliana*. International Journal of Molecular Sciences. 2013;**14**:4545-4559

[23] White PJ, Broadley MR. Mechanisms of caesium uptake by plants. The New Phytologist. 2000;**147**:241-256

[24] Hampton CR, Broadley NR, White PJ. Short review: The mechanisms of radiocaesium uptake by *Arabidopsis* roots. Nukleonika. 2005;**50**:S3-S8

[25] Remy E, Cabrito TR, Batista RA, Teixeira MC, Sá-Correia I, Duque P. The major facilitator superfamily transporter ZIFL2 modulates cesium and potassium homeostasis in *Arabidopsis*. Plant & Cell Physiology. 2015;**56**:148-162

[26] Qi Z, Hampton CR, Shin R, Barkla BJ, White PJ, Schachtman DP. The high affinity K+ transporter AtHAK5 plays a physiological role in planta at very low K+ concentrations and provides a caesium uptake pathway in *Arabidopsis*. Journal of Experimental Botany. 2008;**59**:595-607

[27] Ruiz-Lau N, Bojórquez-Quintal E, Benito B, Echevarría-Machado I, Sánchez-Cach LA, Medina-Lara MF, et al. Molecular cloning and functional analysis of a Na+-insensitive K+ transporter of *Capsicum chinense* Jacq. Frontiers in Plant Science. 2016;**7**:1980. DOI: 10.3389/fpls.2016.01980

[28] Cabrera WE, Schrooten I, De Broe ME, D'Haese PC. Strontium and bone. Journal of Bone and Mineral Research. 1999;**14**:661-668

[29] Smith JT, Sasina NV, Kryshev AI, Belova NV, Kudelsky AV. A review and test of predictive models for the bioaccumulation of radiostrontium in fish. Journal of Environmental

Radioactivity. 2009;**100**:950-954. DOI: 10.1016/j.jenvrad.2009.07.005

[30] White PJ, Broadley MR. Calcium in plants. Annals of Botany. 2003;**92**:487-511

[31] Kellett GL. Alternative perspective on intestinal calcium absorption: Proposed complementary actions of Ca(v)1.3 and TRPV6. Nutrition Reviews. 2011;**69**:347-370. DOI: 10.1111/j.1753-4887.2011.00395.x

[32] Pérez AV, Picotto G, Carpentieri AR, Rivoira MA, Peralta López ME, Tolosa de Talamoni NG. Minireview on regulation of intestinal calcium absorption. Emphasis on molecular mechanisms of transcellular pathway. Digestion. 2008;**777**:22-34. DOI: 10.1159/000116623.

[33] He P, Hou X, Aldahan A, Possnert G, Yi P. Iodine isotopes species fingerprinting environmental conditions in surface water along the northeastern Atlantic Ocean. Scientific Reports. 2013;**3**:2685. DOI: 10.1038/srep02685

[34] Muramatsu Y, Matsuzaki H, Toyama C, Ohno T. Analysis of ^{129}I in the soils of Fukushima prefecture: Preliminary reconstruction of ^{131}I deposition related to the accident at Fukushima Daiichi nuclear power plant (FDNPP). Journal of Environmental Radioactivity. 2015;**139**:344-350. DOI: 10.1016/j.aca.2008.11.013.

[35] Carvalho DP, Dupuy C. Thyroid hormone biosynthesis and release. Molecular and Cellular Endocrinology. 2017;**15**(458):6-15. DOI: 10.1016/j.mce

[36] Küpper F, Schweigert N, Ar Gall E, Legendre JM, Vilter H, Kloareg B. Iodine uptake in *Laminariales* involves extracellular, haloperoxidase-mediated oxidation of iodide. Planta. 1998;**207**:163-171

[37] Dushenkov S, Vasudev D, Kapulnik Y, Gleba D, Fleisher D, Ting KC, et al. Removal of uranium from water using terrestrial plants. Environmental Science & Technology. 1997;**31**:3468-3474. DOI: 10.1021/es970220l

[38] Broadley MR, Willey NJ, Mead A. A method to asses taxonomic variation in shoot caesium concentration among flowering plants. Environmental Pollution. 1999;**106**:341-349

[39] Watanabe T, Broadley MR, Jansen S, White PJ, Takada J, Satake K, et al. Evolutionary control of leaf element composition in plants. The New Phytologist. 2007;**174**:516-523

[40] Sasmaz A, Sasmaz M. The phytoremediation potential for strontium of indigenous plants growing in a mining area. Environmental and Experimental Botany. 2009;**67**:139-144

[41] Fuhrmann M, Lasat MM, Ebbs SD, Kochian LV, Cornish J. Uptake of cesium-137 and strontium-90 from contaminated soil by three plant species; application to phytoremediation. Journal of Environmental Quality. 2002;**31**:904-909

[42] Watanabe I, Tensho K. Further study on iodine toxicity in relation to "reclamation Akagare" disease of lowland rice. Soil Science & Plant Nutrition. 1970;**16**:192-194

[43] Kobayashi D, Okouchi T, Yamagami M, Shinano T. Verification of radiocesium decontamination from farmlands by plants in Fukushima. Journal of Plant Research. 2014;**127**:51-56. DOI: 10.1007/s10265-013-0607-x

[44] MAFF (Ministry of Agriculture, Forestry and Fisheries). About a Removal Technology of Radionuclide (Decontamination Technology) from the Farmland Soil. Press Release 14 September 2011 (in Japanese)

[45] Phang SM, Chu WL, Rabiei R. Phycoremediation. In: Sahoo D, Seckbach J, editors. The Algae World. Dordrecht: Springer Netherlands; 2015. pp. 357-389. DOI: 10.1007/978-94-017-7321-8

[46] Fukuda S, Iwamoto K, Atsumi M, Yokoyama A, Nakayama T, Ishida K, et al. Global searches for microalgae and aquatic plants that can eliminate radioactive cesium, iodine and strontium from the radio-polluted aquatic environment: A bioremediation strategy. Journal of Plant Research. 2014;**127**:79-89. DOI: 10.1007/s10265-013-0596-9

[47] Araie H, Sakamoto K, Suzuki I, Shiraiwa Y. Characterization of the selenite uptake mechanism in the coccolithophore *Emiliania huxleyi* (Haptophyta). Plant & Cell Physiology. 2011;**52**:1204-1210. DOI: 10.1093/pcp/pcr070

[48] Iwamoto K, Shiraiwa Y. Characterization of intracellular iodine accumulation by iodine-tolerant microalgae. Procedia Environmental Sciences. 2012;**15**:34-42. DOI: 10.1016/j.proenv.2012.05.007

[49] Bardi U. Extracting minerals from seawater: An energy analysis. Sustainability. 2010;**2**:980-992. DOI: 10.3390/su2040980

[50] Nakayama T, Nakamura A, Yokoyama A, Shiratori T, Inouye I, Ishida K. Taxonomic study of a new eustigmatophycean alga, Vacuoliviride crystalliferum gen. et sp. nov. Journal of Plant Research. 2015;**128**:249-257. DOI: 10.1007/s10265-014-0686-3

[51] Singh S, Thorat V, Kaushik CP, Raj K, Eapen S, D'Souza SF. Potential of *Chromolaena odorata* for phytoremediation of 137Cs from solution and low level nuclear waste. Journal of Hazardous Materials. 2009;**162**:743-745. DOI: 10.1016/j.jhazmat.2008.05.097

[52] Fukuda S, Iwamoto K, Yamamoto R, Minoda A. Cellular accumulation of cesium in the unicellular red alga *Galdieria sulphuraria* under mixotrophic conditions. Journal of Applied Phycology. In press. DOI: 10.1007/s10811-018-1525-z

[53] Honda M, Shimoyama I, Kogura T, Baba Y, Suzuki S, Yaita T. Proposed cesium-free mineralization method for soil decontamination: Demonstration of cesium removal from weathered biotite. ACS Omega. 2017;**2**:8678-8681. DOI: 10.1021/acsomega.7b01304

Chapter 2

Cyanobacteria Growth Kinetics

Leda Giannuzzi

Abstract

Harmful cyanobacterial blooms are a global problem for freshwater ecosystems used for drinking water supply and recreational purposes. Cyanobacteria also produce a wide variety of toxic secondary metabolites, called cyanotoxins. High water temperatures have been known to lead to cyanobacterial bloom development in temperate and semiarid regions. Increased temperatures as a result of climate change could therefore favor the growth of cyanobacteria, thus augmenting the risks associated with the blooms. Though temperature is the main factor affecting the growth kinetics of bacteria, the availability of nutrients such as nitrogen and phosphorus also plays a significant role. This chapter studies the growth kinetics of toxin-producing *Microcystis aeruginosa* and evaluates potential risks to the population in scenarios of climate change and the presence of nutrients. The most suitable control methods for mitigation are also evaluated.

Keywords: modeling of cyanobacterial, harmful cyanobacterial, *Microcystis aeruginosa*, growth kinetics, control of harmful cyanobacterial

1. Introduction

Eutrophication resulting from harmful cyanobacterial blooms is a frequent nuisance phenomenon in freshwater lakes and estuaries around the world, posing a serious threat to aquatic ecosystems and human health [1, 2]. Cyanobacteria thus constitute a global problem in freshwater ecosystems used for drinking water and recreational purposes [3]. The potential damage to water supplies, recreation, tourism, aquaculture, and agriculture could have a substantial economic and social impact. The most commonly occurring genera of cyanobacteria include *Microcystis, Oscillatoria, Anabaena*, and *Aphanizomenon*.

For well over a century, many animal and human poisonings have been associated with Cyanobacteria and their toxins; the death of livestock, wildlife, and pets due to ingestion of water containing toxic cyanobacterial cells or toxins released by the cells has been extensively documented. Human poisonings have also been reported [4]. The occurrence of toxic cyanobacteria has become a worldwide problem [5, 6].

One group of toxic compounds synthesized by several cyanobacteria (*Microcystis, Anabaena, Planktothrix,* and *Nostoc*) comprises numerous hepatotoxic cyclic heptapeptide microcystins [7, 8].

In Argentina, in recent decades, blooms have been recorded in rivers, lakes, coastal lagoons, and estuaries throughout the country, demonstrating the geographical extent of the problem. An increase has been detected both in the number of responsible species and in the frequency and intensity of the harmful events. The genera most commonly associated with the development of toxic blooms are *Microcystis* and *Dolichospermum* (ex *Anabaena*), and the most cited cyanotoxins are microcystins [9–11]. In 2014, a series of harmful episodes caused by cyanobacteria blooms associated with water treatment systems in different parts of Argentina occurred. The presence of cyanobacteria and cyanotoxins has been reported in several sources of drinking water. *Microcystis* colony cells and microcystins were detected in water in the cities of La Plata and Ensenada, Buenos Aires, Argentina, in 2006 [12], evidencing the inefficiency of the water treatment plant.

Giannuzzi et al. [13] reported an acute case of cyanobacterial poisoning in the Salto Grande dam, Argentina, which occurred in January 2007. A young man accidentally became immersed in an intense bloom of *Microcystis* spp. with 48.6 µg L^{-1} of microcystin-LR in water samples. The patient was hospitalized in intensive care and diagnosed with an atypical pneumonia. A week after exposure, the patient developed hepatotoxicosis with a significant increase in hepatic damage biomarkers (ALT, AST, and γGT). Complete recovery took 20 days. It is not known whether there was an eventual chronic intoxication after the acute poisoning. In the year 2000 in Bahía Blanca (Buenos Aires, Argentina), alterations were detected in the organoleptic characteristics of the water network (unpleasant odor and taste), product of the liberation of geosmin by *Dolichospermum circinalis* blooms. This episode coincided with the appearance of dermal and respiratory problems in the population [14].

The duration of cyanobacterial blooms in temperate zones can last 2–4 months during the summer period, whereas in tropical and subtropical regions of Australia, China, and Brazil, they can sometimes persist all year round [15].

The major factors that influence the growth of cyanobacteria are light, temperature, and the nutrients composition of the suspending medium.

High water temperatures have been known to lead to cyanobacterial bloom development in temperate [16–18] and semiarid regions [19]. Increasing air and water temperatures as a result of climate change are likely to promote a faster algal growth rate [20, 21].

Nitrogen (N) and/or phosphorus (P) levels can also positively affect cyanobacterial growth in lakes and river. The absolute and relative concentrations of these nutrients affect the growth rate, abundance, and composition of phytoplankton in lake water [22] as commonly measured in terms of their trophic state, defined as the total weight of biomass in a given water body at the time of measurement [23]. Many studies show that phosphorus is the limiting nutrient in freshwater bodies [24, 25], and other studies show the relationship between cyanobacterial abundance and phosphorus concentrations in lakes [26, 27].

The trophic state of a lake generally increases with increases in total nitrogen (TN) and total phosphorus (TP) concentrations. Resolving lake or river eutrophication problems calls for a better understanding of the water and air temperature-dependence of algal blooms.

A high P concentration is considered to be the main cause of *Microcystis* blooms in the Nakdong River of Korea [28, 29]. Schindler [30, 31] report that N is unlikely to be the limiting factor for blooms because of the presence of N_2-fixing cyanobacterium in water bodies. Phosphate (PO_4^{3-}) is released from the sediment during summer, absorbed by *Microcystis* and stored in the bottom layer [32, 33]; using its gas vacuole, the *Microcystis* then moves toward the high-intensity light at the surface to generate blooms [34–36].

Provided factors such as illumination and nutrients remain saturating, and the photosynthetic and specific maximal growth rate responses of different algal species to temperature can be compared [34].

Physiological properties within a single species, including photosynthetic response, can change according to the growth conditions [37]. Photoperiodicity- and light intensity-dependent changes in photosynthetic parameters and different pigments such as chlorophyll a and phycocyanin are to be expected.

The general consensus is that the optimum growth temperature for cyanobacteria is higher than that for most algae. Paerl [38] reported the optimum temperature to be higher than 25°C, overlapping with that of green algae (27–32.8°C) but clearly differing from that of dinoflagellates (17–27°C) and diatoms (17–22°C). Crettaz Minaglia [39] reported the optimum growth temperature for native *M. aeruginosa* to be 33.39 ± 0.55°C.

Lürling [40] found similar optimal temperatures for two strains of *M. aeruginosa* (30.0–32.5°C). These data suggest that the native strain of *M. aeruginosa* is able to compete favorably with other phytoplankton species, producing more frequent blooming events in scenarios of climate change.

Paerl [41] reported that higher temperatures (up to 25°C) due to climate change may lead to increased cyanobacterial growth rates and thus higher cyanobacterial dominance in temperate water bodies [17, 20]. This trend would be further facilitated by cyanobacterial buoyancy, which aids their proliferation in increasingly stratified conditions because decreasing water viscosity at higher temperatures results in higher flotation velocities of buoyant cyanobacteria [19, 42].

Many authors describe an inverse relationship between temperature and microcystin production.

Crettaz Minaglia [39] found that the production of MC-LR decreased with increasing temperature, coinciding with the findings of [43–50].

van der Westhuizen [51] reported that optimal growth conditions do not coincide with the production of toxins. Similarly, Gorham [43] affirmed that the optimum temperature for growth (30–35°C) differed from that for optimal toxicity (25°C).

In an interesting paper, Mowe et al. [52] suggest that higher mean water temperatures resulting from climate change will generally not affect *Microcystis* spp. growth rates in Singapore, except for increases in *M. ichthyoblabe* strains. However, depending on the species, the toxin cell ratio may increase under moderate warming scenarios.

Further studies on the temperature dependency of the different physiological processes affecting growth (e.g., carbon fixation, photorespiration, and respiration) are required in order to better understand the differences in temperature sensitivity between *Microcystis* growth and toxins production.

2. Modeling *M. aeruginosa* growth

An evaluation of microbiological cyanobacterial processes calls for kinetics studies examining the rates of production of cells and their metabolites and the effects of various factors on these rates.

One of the basic tools in microbiology is growth kinetics, defined as the relationship between a specific growth rate and parameters such as temperature, pH, light intensity, short wavelength radiations, pH, and nutrients.

A convenient way to evaluate laboratory-based bacteria growth systems under different abiotic factors is to examine the parameters characterizing the three phases of bacterial growth: the lag phase, the exponential phase, and the stationary phase.

During the lag phase, which can last from 1 hour up to several days, there is very little change in the number of bacteria cells because while they are adapting to the growth conditions, they are still immature and unable to reproduce. This is the period when the synthesis of RNA, enzymes, and other molecules occurs.

The exponential phase is characterized by cell doubling. The number of new bacteria appearing per unit time is proportional to the present population. With no limitations in place, doubling continues at a constant rate, leading to a doubling of the number of cells and the rate of population increase with each consecutive time period. Plotting the logarithm of cell number against time produces a straight line, the slope of which indicates the specific growth rate of the organism, which is a measure of the number of divisions per cell per unit time. The actual rate of this growth depends on the growth conditions, which affect the frequency of cell division events and the probability of both daughter cells surviving. Under controlled conditions, the cyanobacteria population can be doubled four times a day and then tripled. However, this exponential growth eventually comes to an end when the medium becomes depleted of nutrients and enriched with waste.

The stationary phase results when the death rate is equal to the growth rate, often because of the depletion of an essential nutrient and/or the formation of an inhibitory product such as an organic acid, giving rise to a "smooth," horizontal line on the curve.

The final phase is the death phase. Bacterial death can be the result of lack of nutrients, environmental temperature above or below the tolerance band for the species, or other deleterious conditions.

Modeling a cyanobacterial growth curve allows one to reduce recorded data to a limited number of parameters of interest such as the specific growth rate, lag phase duration, and maximum population density.

The growth models found in the literature describe only the number of organisms and do not include substrate consumption as would a model based on the Monod equation. However, the substrate level is not of interest in our application since we assume there to be sufficient substrate to allow cyanobacterial growth.

An assessment of natural populations of *Microcystis aeruginosa* requires data on pure culture growth under well-defined conditions.

Studying the growth kinetics of *Microcystis* in relation to nutrient concentrations is very important for management purposes [53].

In batch culture methods, the culture is not maintained at a specific growth stage with constant addition and removal of culture medium and cells [54].

This makes it an appropriate method for our *M. aeruginosa* study since the natural ecosystem is not steady state either: weather-related factors cause changes in nutrient loading, resulting in varying nutrient concentrations, with no expected resupply of nutrients to the water column. *Microcystis* habitats are therefore more like batch experiments than continuous cultures able to reach a steady state [55].

The basic batch culture growth model emphasizes aspects of bacterial growth, which may differ from the growth of other organisms. Plotting an experimentally determined cell number or cell mass concentration (or their logarithms) against time gives rise to a characteristic curve.

For the purpose of modeling the growth of *M. aeruginosa*, three primary continuous population models can be used.

The simplest is a linear function based on the exponential or Malthus model, called a simple exponential growth model. This model assumes that the growth rate of the population is proportional to its density

$$lnN \ = \ \ln(N_0) + \mu * t \tag{1}$$

where N(t) is the population at time t, N_0 is the initial population, μ is a constant indicating the rate of growth, and t is time. The parameter μ is called the specific growth rate and is expressed in reciprocal time units.

This model is widely used in microorganisms and is also very useful for describing the population growth of many organisms over short periods of time, there being no space or resource limitations.

During a batch cultivation, the specific growth rate changes continuously from zero to a maximum value, μ_{max}. The value of the maximum specific growth rate depends on the type of microorganism and on physical and chemical cultivation conditions (temperature, pH, medium composition, light, etc.). Under given culture conditions, it is constant and represents an important characteristic of the process.

The specific growth rate (μ) can be calculated between successive sampling points from a simple first-order rate law using the equation

$$\mu = \frac{ln\frac{N_1}{N_0}}{t_1 - t_0} \tag{2}$$

where N_0 is the cell number per mL culture at time t_0 and N_1 is the cell number at time t_1.

The main parameter characterizing the growth rate is the specific growth rate, which can be used to express other growth parameters given below.

The relationship between the specific growth rate (μ) and cell number doubling time (t_d) can be obtained by inserting into the equation $N = 2N_0$ and $t = t_d$.

$$t_d = \frac{ln2}{\mu} = \frac{0.693}{\mu} \tag{3}$$

Yet another model is the logistic or Verhulst model (Eq. (4)), a quadratic function based on the previous model under the assumption that the population cannot grow indefinitely and faster. In this model, μ is not a constant, but is a linearly increasing function of population density. This model has two equilibrium points defined as N values where the growth velocity is zero. These points correspond to $N = 0$ and $N = K$ (load capacity). The load capacity refers to the maximum population that its environment can sustain in terms of resource or space availability.

$$N(t) = \frac{K}{1 + \frac{K - N_0}{N_0}e^{-\mu.t}} \tag{4}$$

where N(t) is the final population, N_0 is the initial population, μ is a constant that indicates the growth rate, t is time, and K is the load capacity.

Growth rate (μ) is commonly expressed in the literature as a function of light, nutrient, pH, ionic conditions, and temperature. Modeling the growth rate is based on simply multiplying the functions upon which growth is dependent:

$$\mu = f(N) * f(I) * f(T) \tag{5}$$

where μ (time^{-1}) is the cyanobacterial growth rate and f(I), f(N), and f(T) are the effects of irradiance, nutrients, and temperature on the growth rate, respectively.

Another method to calculate the three parameters of the three phases of bacterial growth mentioned earlier is using the modified Gompertz equation, a double exponential function based on four parameters, which describes an asymmetric sigmoid curve Eq. (6) [56].

The Gompertz model is one of the most widely used and recommended models from which lag time, maximum growth rate, and maximum population density (stationary phase) can be obtained directly from nonlinear regression of the cell numbers versus time data [57, 58].

A Gompertz model describing the growth of *M. aeruginosa* would be a good fit for analyzing the effect of temperature, irradiance, and nutrients on the parameters of kinetic growth curves in batch culture.

$$log(N) = a + c \times exp(-exp(-b \times (t - m))) \qquad (6)$$

where log(N) is the decimal logarithm of the cell counts (log (cell mL^{-1}), t is time (days), and a is the logarithm of the asymptotic counts when time decreases indefinitely (roughly equivalent to the logarithm of the initial levels of cyanobacteria (log (cell mL^{-1})), c is the logarithm of the asymptotic counts when time is increased indefinitely (the number of log cycles of growth) (log (cell mL^{-1})), b is the growth rate relative to time (days^{-1}), and m is the required time to reach the maximum growth rate (days).

The maximum or specific growth rate (μ) value is defined as the tangent in the inflection point and was calculated as μ = b.c/e with e = 2.7182 (days^{-1}). The lag phase duration (LPD) is defined as the t-axis intercept of this tangent, the asymptote, and was calculated as LPD = m − 1/b, (days); the maximum population density MPD = a + c (log (cell mL^{-1}) was derived from these parameters [59].

The value of modeling has been recognized for a number of years. Accurate and well-validated models are able to predict the behavior of dynamically changing systems and provide data and insights that would be difficult or impossible to obtain by conventional field.

Figure 1 shows a detail of *M. aeruginosa* isolated from the environment. When these cells are grown in culture medium, they lose their capacity to colonies form. In this review, the Gompertz equation was applied to *M. aeruginosa* growth in culture media (BG11) at different temperatures. The cells mL^{-1} data were fitted to the Gompertz equation by nonlinear regression using the program Systat (Systat Inc., version 5.0). The selected algorithm calculates the set of parameters with the lowest residual sum of squares and a 95% confidence interval for *M. aeruginosa* growth.

In a previous work, we informed two additional kinetic parameters: generation time (GT) and the relative lag phase duration (RLPD). Generation time was defined as the time for the bacterial population to double in cell numbers and was calculated by dividing μ values by 0.301 (equivalent to log$_{10}$ 2); GT is thus a measure of the metabolic rate in a new environment. The RLPD, defined by the amount of work to be done in adjusting to a new environment and the rate at which that work is done, was calculated by dividing LPD by GT [39].

For *M. aeruginosa*, the growth rate shows a lag phase followed by an exponential phase and finally a decreasing growth rate down to zero, resulting in a maximum value of the number of cells (**Figure 2**). During the experiment, the number of cells mL^{-1} increased exponentially in all cultures after a lag phase. **Figure 2** shows *M. aeruginosa* counts growing in BG11 modified medium at 26, 28, 30, and 35°C.

By examining these parameters, the blooming behavior of *M. aeruginosa* can be predicted and early warning signs recognized in time to take preventive action. Although some authors have performed laboratory experiments using *M. aeruginosa* under different temperature conditions, irradiance and N:P ratio [51, 60–64], and modeled their growth, they only applied the linear growth model to the exponential phase of the curve.

The Gompertz parameters of these curves were reported by Crettaz Minaglia et al. [39]. It can be seen that as the temperature increases, the value of μ increases and LPD decreases. Thus, when the temperature changed from 26 to 35°C, the μ values increased from 0.18 to 0.24 days^{-1}, and LPD decreased from 4.10 to 0.75 days, nonsignificant differences were found for the MPD values (7.25–7.10 log CFU mL^{-1}). The GT parameter ranged from 1.67 to 1.25 days, and RLPD from 2.40 to 0.60.

The ratio of the specific growth rate to the lag time duration was approximately constant, suggesting a linear relationship between lag phase and the reciprocal of the growth rate. This finding was corroborated by Crettaz Minaglia [39], who reported that the lag phase showed a linear behavior with the reciprocal specific growth rate for *M. aeruginosa*. The correlation coefficient (R^2) was 0.86.

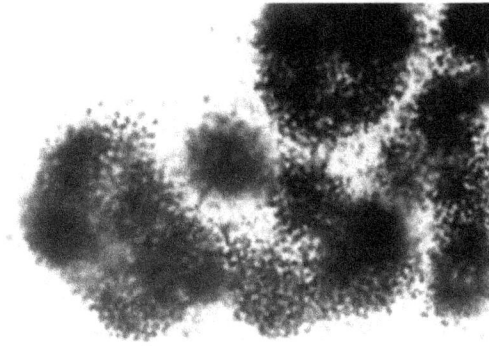

Figure 1.
Light microscopy image of a Microcystis aeruginosa colony isolated from the environment (kindness of Ricardo Echenique).

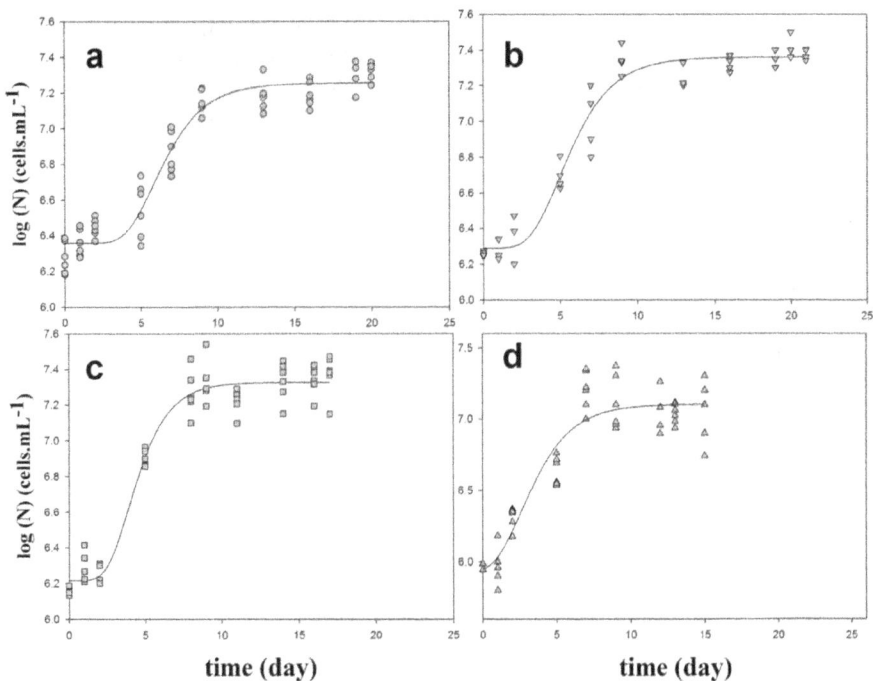

Figure 2.
Effect of temperature on M. aeruginosa growth in culture media: (a) 25, (b) 28, (c) 30, and (d) 36°C.

Lyck [61] reported specific growth rate values ranging from 0.52 to 0.54 day^{-1} calculated between successive sampling times according to a simple first-order rate function using cell concentration (cells mL^{-1}).

In reviewing available literature on the effects of temperature on growth rates, Canale and Vogel [65] concluded that as temperature increased, the highest growth rates for broad phytoplankton groups changed from diatoms, via green algae to cyanobacteria (blue-green algae). Species-specific responses are, however, highly variable [66].

The specific growth rate and lag phase duration are known to be affected by many variables, and the cyanobacterial responses to changes in the environment are complex and difficult to characterize. **Figure 3** shows an example of the variation

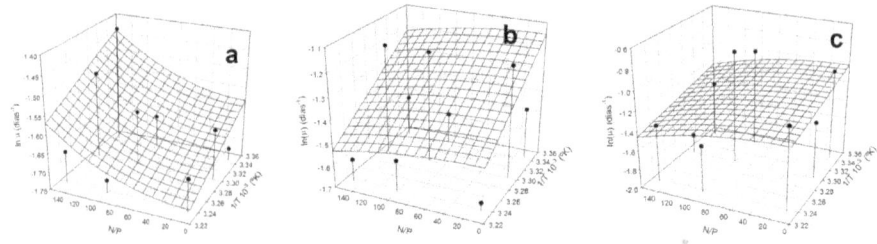

Figure 3.
Surface response plots showing the dependence of specific growth rate parameters of M. aeruginosa on temperature and N/P ratio: (a) 30, (b) 50, and (c) 70 μmol photons $m^{-2} s^{-1}$.

in the specific growth rate with changes in temperature (26, 30, and 36°C), nutrient (N/P 10, 100, and 150), and irradiance (30, 50, 70 μmol photon $m^{-2} s^{-1}$) conditions for *M. aeruginosa* growing in culture medium. For combined temperature and the N/P ratio, we chose the Arrhenius-type temperature dependence model for each irradiance as the starting point for developing a model including both temperature and N/P effect. After fitting to different models, those with the highest correlation coefficients and the lowest errors in the estimated parameters were selected. The following Eq. (7) was obtained by stepwise analysis with statistical SYSTAT software and describes both the inverse absolute temperature effect (25, 30, and 36°C) and N/P ratio dependence (10, 100, and 150) on the specific growth rate.

$$ln\mu = K1 + K2 * \frac{1}{T} + K3 * \left(\frac{N^2}{P}\right) \tag{7}$$

The percent variance was very high at 98.8%, indicating a very good fit of the model to the data. The parameter K_2 was 0.89, 2.25, and 1.56 for 30, 50, and 70 μmol photons $m^{-2} s^{-1}$, respectively. The values of $K_2 = E_a/R$ where E_a is the activation energy of μ (KJ mol^{-1}) applied Eq. (7) and R is the gas constant (8.31 KJ $K^{-1} mol^{-1}$). In the present study, E_a was 7.40, 18.69, and 12.96 KJ $K^{-1} mol^{-1}$ for 30, 50, and 70 μmol photons $m^{-2} s^{-1}$, respectively.

Figure 3a–c shows examples of a surface response plot corresponding to Eq. (7) obtained by fitting ln μ of *M. aeruginosa* versus temperature and N/P ratio.

Using the model reported here, we determined the combined effects of the N/P ratio and temperature on specific growth rates in controlled laboratory assays, thus enabling us to predict *M. aeruginosa* growth under different conditions from those tested experimentally in this work, but within the studied range of temperatures and N/P ratio.

The Gompertz model was successfully tested with the experimental data for *M. aeruginosa* at different temperature, ration N/P, and light intensity (data not show). It is very important to test the model under different conditions or to verify the model for other species of cyanobacteria and diatoms.

However, many open questions remain concerning the validity of applying laboratory-observed growth kinetics to environmental growth conditions, with diverging data being reported for pure cultures growing with single substrates.

Although our model only takes into account temperature and N/P ratio, it would be important to extend the modeling to other factors such as pH and elements such as metals (Fe, Mn, etc.). Further studies are required to gain deeper insight into the factors that influence growth in order to better predict aspects related to *M. aeruginosa* blooms.

3. Cyanobacterial control

Cyanobacterial blooms can lead to the accumulation of cyanotoxins in aquatic animals, eventually posing a high risk to human health as well.

In the current scenario of growing problems associated with cyanobacterial blooms and their toxins, an environmentally compatible control strategy is urgently required. The removal of harmful cyanobacterial blooms is a crucial step for the adequate maintenance of water supplies and for the safety of food and aquatic products. Controlling cyanobacterial blooms is likely to become an even more challenging task in the future due to global warming effects.

Despite the availability of control methods for cyanobacterial blooms, it has not yet been possible to prevent the excessive proliferation of these organisms, which have adapted so successfully to water surfaces. The effectiveness of control methods naturally varies according to the circumstances (type and size of the lake, retention time, degree of alteration, quantity of nutrient load, quality and quantity of sediments, season, amount of aquatic life, etc.); they are not universal and their use may be restricted to particular circumstances.

The preferred method for preventing these blooms is to reduce the availability of nutrients, especially phosphorus, the main cause of the massive presence of cyanobacteria. This implies the rehabilitation of point and nonpoint sources of nutrients (discharge of effluents, drift of chemical substances from agriculture, and erosion of urban and forest areas) [67]. In those cases where nutrient reduction is not possible, more drastic, short-term action has been proposed in the form of chemical, physical, and biological approaches [68], each with its advantages and disadvantages for application to the control of harmful algal blooms.

A widely adopted chemical approach is the addition of algaecide (copper sulfate), oxidants (chlorine, potassium permanganate), and flocculants ($FeCl_3$, $AlCl_3$, polyaluminum chloride) etc., all of which have proven to be efficient in removing cyanobacteria cells. However, though chemical approaches can take rapid effect in removing algal blooms, they can cause secondary pollution of aquatic environments [69]. Their main disadvantage is that they do not selectively target harmful cyanobacteria and can lead to the elimination or damage to nonharmful algae or beneficial organisms. Depending on the oxidant and cyanotoxin type, some oxidants can cause the release of toxins, and the subsequent rapid oxidation of the toxins must therefore be assured [70].

The application of chemical agents to lakes and water bodies often leads to the collapse of aquatic ecosystems.

Hydrogen peroxide (HP) is selective for cyanobacteria (vs. eukaryotic algae and higher plants) and poses no serious long-term threats to the system because of its rapid decomposition without producing persistent toxic chemicals or by-products that cause esthetic odor or color issues. It has been reported that HP has potential for removing *Microcystis* sp. and microcystins in different environments. Lakes dominated by *M. aeruginosa*, *Aphanizomenon*, and *Dolichospermum* (formerly called Anabaena) have been successfully treated with HP [71]. It is important to assess the impact of HP on elements of the ecosystem such as larval fish, macroinvertebrates, and zooplankton.

Physical approaches, such as mixing lake waters using an air compressor, ultrasonic damage to algal cells and pressure devices to collapse cyanobacterial gas vesicles, have also been proposed to control algal blooms. Other treatments such as the mechanical removal of cyanobacterial biomass and sediments and hypolimnetic aeration and oxygenation have also been described.

The most apparent merit of physical approaches for the removal of algae as opposed to chemical manipulations is that they are less likely to give rise to secondary pollution. However, the physical removal of algae is energy intensive and tends to be of low efficiency. Moreover, injury to nontarget organisms by energy-intensive treatments also limits the field application on a large scale [72].

Though biological approaches to controlling toxic cyanobacteria and harmful cyanobacterial blooms tend to be environmentally friendly, their efficiency is determined by many biotic and abiotic factors in the environment. It is well known that MCs can be degraded by local bacterial communities frequently exposed to cyanobacterial blooms.

The removal of MCs has been reported by a group of microorganisms generically referred to as a consortium [73].

Furthermore, a large group of bacteria able to degrade MCs has been isolated, Sphingomonadaceae being the most studied family. Most of these organisms have been identified as *Sphingomonas* [74] and *Sphingopyxis* [75].

Some biologically derived bioactive substances inhibit the growth of aquatic bloom-forming cyanobacteria [76–78], including plant extracts and identified natural chemicals from plants and microorganisms.

Aquatic plants such as *Stratiotes aloides* [79] *Myriophyllum spicatum* [80], *Phragmites communis* [81], *Ceratophyllum demersum* and *Najas marina* spp., *Intermedia* [82], and extracts of *Ephedra equisetina* root have been reported to inhibit the growth of cyanobacteria. Most of these substances are biodegraded in natural environments. However, actual field applications to control harmful cyanobacteria are currently very limited owing to the high cost of algicide preparations and low algae-removal efficiency compared to chemical algicides.

In view of the paucity of studies on the ecological and public health risks associated with most antialgal substances, their application should be very carefully evaluated. Only ecologically safe and easily applicable substances should be used for cyanobacterial growth control.

4. Conclusions

Mathematical modeling applied to *M. aeruginosa* growth is an efficient tool to predict the effect of different variable as temperature, irradiance, and nutrients on the kinetic parameters. The current study provides quantitative evidence of the effects of temperature, irradiance, and nutrients on *M. aeruginosa* growth. The above data suggest that the native strain of *M. aeruginosa* is able to compete favorably with other phytoplankton species, producing more frequent blooming events in scenarios of climate change. In the current scenario of growing problems associated with cyanobacterial blooms and their toxins, an environmentally compatible control strategy is urgently required. However, the use of control agents, whether physical, chemical, and biological, is not yet sufficiently safe due to certain harmful effects on the environment. Only ecologically safe and easily applicable substances should be used for cyanobacterial growth control.

Acknowledgements

This study was financially supported by University National of La Plata (UNLP X646), National Agency of Scientific and Technical Research (PICT0861-2013), and CONICET (PIP0959).

Conflict of interest

No conflict of interest.

Author details

Leda Giannuzzi
Department of Biological Science, University of La Plata, Centro de Investigacion y
Desarrollo en Criotecnologia de Alimentos (CIDCA), La Plata, Argentina

*Address all correspondence to: leda@biol.unlp.edu.ar

IntechOpen

References

[1] Hudnell HK. The state of U.S. freshwater harmful algal blooms assessments, policy and legislation. Toxicon. 2010;**55**:1024-1034. DOI: 10.1016/j.toxicon.2009.07.021

[2] Carey CC, Ibelings BW, Hoffmann EP, Hamilton DP, Brookes JD. Eco-physiological adaptations that favour freshwater cyanobacteria in a changing climate. Water Research. 2012;**46**:1394-1407. DOI: 10.1016/j.watres.2011.12.016

[3] Paerl HW, Fulton RS, Moisander PH, Dyble J. Harmful freshwater algal blooms, with an emphasis on cyanobacteria. The Scientific World. 2001;**1**:76-113. DOI: 10.1100/tsw.2001.16

[4] Azevedo S, Carmichael WW, Jochimsen E, Rinehart K, Lau S, Shaw G, et al. Human intoxication by microcystins during renal dialysis treatment in Caruaru/Brazil. Toxicology. 2002;**181/182**:441-446

[5] Codd GA. Cyanobacterial toxins: Occurrence, properties and biological significance. Water Science and Technology. 1995;**32**:149-156

[6] Namikoshi M, Rinehart KL. Bioactive compounds produced by cyanobacteria. Journal of Industrial Microbiology. 1996;**17**:373-384. DOI: 10.1007/bf01574768

[7] Carmichael WW. Cyanobacteria secondary metabolites-the cyanotoxins. A review. The Journal of Applied Bacteriology. 1992;**72**:445-459. DOI: 10.1111/j.1365-2672.1992.tb01858.x

[8] Carmichael WW. The cyanotoxins. Advances in Botanical Research. 1997;**27**:211-256. DOI: 10.1016/s0065-2296(08)60282-7

[9] Echenique RO, Aguilera A. Cyanobacteria toxígenas: aspectos generales para su identificación taxonómica. In: Giannuzzi L, editor. Cianobacterias y cianotoxinas: identificación, toxicología, monitoreo y evaluación de riesgo. Corrientes: Moglia Impresiones; 2009. pp. 37-51

[10] Azevedo S. South and Central America: Toxic cyanobacterial. In: Codd GA, Azevedo SMFO, Bagchi SN, Burch MD, Carmichael WW, Harding WR, Kaya K, Utkilen HC, editors. CYANONET. A Global Network for Cyanobacterial Bloom and Toxin Risk Management. Initial Situation Assessment and Recommendations, Technical documents in Hydrology PHI-VI. 2005. pp. 115-126

[11] Otaño S, Salerno G, Ruiz M, Aguilera A, Echenique RO. ARGENTINA: Cyanobacteria and cyanotoxins: Identification, toxicology, monitoring and risk assessment. In: Chorus I, editor. Current Approaches to Cyanotoxin Risk Assessment, Risk Management and Regulations in Different Countries. Wörlitzer Platz 1, 06844 Dessau-Roßlau, Germany: Federal Environment Agency (Umweltbundesamt); 2012. pp. 16-20

[12] Echenique RO, Aguilera A, Giannuzzi L. Problems on drinking water related to toxigenic cyanobacteria: Some cases studied in Argentina. In: Tell G, Izaguirre I, O'Farrel I, editors. Freshwater phytoplankton from Argentina. Advances in Limnology. Vol. 65. 2014. pp. 431-444. DOI: 10.1127/1612-166X/2014/0065-0055

[13] Giannuzzi L, Sedan D, Echenique R, Andrinolo D. An acute case of intoxication with cyanobacteria and cyanotoxins in recreational water in Salto Grande dam, Argentina. Marine Drugs. 2011;**9**:2164-2175. DOI: 10.3390/md9112164

[14] Echenique R, Giannuzzi L, Ferrari L. Drinking water: Problems related

to water supply in bahía Blanca, Argentina. Acta Toxicológica Argentina. 2006;**14**(2):2-9

[15] Sivonen K, Jones G. Cyanobacterial toxins. In: Chorus I, Bartram J, editors. Toxic Cyanobacteria in Water. A Guide to their Public Health Consequences, Monitoring and Management. London: E & FN Spon, WHO; 1999. pp. 41-112. DOI: 10.4324/9780203478073

[16] Robarts RD, Zohary T. Temperature effects on photosynthetic capacity, respiration, and growth rates of bloom-forming cyanobacteria. New Zealand Journal of Marine and Freshwater Research. 1987;**21**:391-399. DOI: 10.1080/00288330.1987.9516235

[17] Jöhnk KD, Huisman J, Sharples J, Sommeijer B, Visser PM, Stroom JM. Summer heatwaves promote blooms of harmful cyanobacteria. Global Change Biology. 2008;**14**:495-512. DOI: 10.1111/j.1365-2486.2007.01510.x

[18] Wu Y, Li L, Gan N, Zheng L, Ma H, Shan K, et al. Seasonal dynamics of water bloom-forming *Microcystis* morphospecies and the associated extracellular microcystin concentrations in large, shallow, eutrophic Dianchi Lake. Journal of Environmental Sciences. 2014:1921-1929. DOI: 10.1016/j.jes.2014.06.031

[19] Mitrovic SM, Oliver RL, Rees C. Critical flow velocities for the growth and dominance of *Anabaena circinalis* in some turbid freshwater rivers. Freshwater Biology. 2003;**48**:164-174. DOI: 10.1046/j.1365-2427.2003.00957.x

[20] Davis TW, Berry DB, Boyer GL, Gobler CJ. The effects of temperature and nutrients on the growth and dynamics of toxic and non-toxic strains of *Microcystis* during cyanobacteria blooms. Harmful Algae. 2009;**8**:715-725. DOI: 10.1139/f05-239

[21] Jones ML, Shuter BJ, Zhao YM, Stockwell JD. Forecasting effects of climate change on Great Lakes fisheries: Models that link habitat supply to population dynamics can help. Canadian Journal of Fisheries and Aquatic Sciences. 2006;**63**:457-468. DOI: 10.1139/f05-239

[22] Bergström A. The use of TN:TP and DIN:TP ratios as indicators for phytoplankton nutrient limitation in oligotrophic lakes affected by N deposition. Aquatic Sciences. 2010;**72**:277-281. DOI: 10.1007/s00027-010-0132-0

[23] Carlson RE. A trophic state index for lakes. Limnology and Oceanography. 1977;**22**:361-369. DOI: 10.4319/lo.1977.22.2.0361

[24] Schindler DW, Hecky RE, Findlay DL, Stainton MP, Parker BR, Paterson MJ, et al. Eutrophication of lakes cannot be controlled by reducing nitrogen input: Results of a 37-year whole-ecosystem experiment. Proceedings of the National Academy of Sciences. 2008;**105**(32):11254-11258. DOI: 10.1073/pnas.0805108105

[25] Saxton MA, Arnold RJ, Bourbonniere RA, McKay RML, Wilhelm SW. Plasticity of total and intracellular phosphorus quotas in *Microcystis aeruginosa* cultures and Lake Erie algal assemblages. Frontiers in Microbiology. 2012;**3**:19. DOI: 10.3389/fmicb.2012.00003

[26] Chaffin JD, Bridgeman TB, Heckathorn SA, Mishra S. Assessment of *Microcystis* growth rate potential and nutrient status across a trophic gradient in western Lake Erie. Journal of Great Lakes Research. 2011;**37**(1):92-100. DOI: 10.1016/j.jglr.2010.11.016

[27] Downing JA, Watson SB, McCauley E. Predicting cyanobacteria dominance in lakes. Canadian Journal of Fisheries and Aquatic Sciences. 2001;**58**(10):1905-1908. DOI: 10.1139/f01-143

[28] Kim EH, Kang SK. The effect of heavy metal ions on the growth of *Microcystis aeruginosa*. Journal of Korean Society on Water Quality. 1993;**9**:193-200. DOI: 10.1016/0041-0101(91)90254-o

[29] Lee TG, Park SW, Yu TS, Kim J. The growth and coagulation characteristics of *Microcystis aeruginosa* during water treatment processes. Journal of the Korean Society of Water and Wastewater. 1998;**6**:33-42

[30] Schindler DW. The dilemma of controlling cultural eutrophication of lakes. Proceedings of the Biological Sciences. 2012;**279**:4322-4333. DOI: 10.1098/rspb.2012.1032

[31] Schindler DW, Hecky RE, Findlay DL, Stainton MP, Parker BR, Paterson MJ, et al. Eutrophication of lakes cannot be controlled by reducing nitrogen input: Results of a 37-year whole-ecosystem experiment. Proceedings of the National Academy of Sciences of the United States of America. 2008;**105**:11254-11258. DOI: 10.1073/pnas.0805108105

[32] Jacobson L, Halmann M. Polyphosphate metabolism in the blue-green alga, *Microcystis aeruginosa*. Journal of Plankton Research. 1982;**4**:481-488. DOI: 10.1093/plankt/4.3.481

[33] Jung H-Y, Cho K-J. Environmental conditions of sediment and bottom waters near sediment in the downstream of the Nagdong River. Korean Journal of Limnology. 2003;**36**:311-321

[34] Reynolds CS, Jaworski GHM, Cmiech HA, Leedale GF. On the annual cycle of the blue-green alga *Microcystis aeruginosa* Kütz. Emend Elenkin. Philosophical Transactions of the Royal Society of London. Series B, Biological Sciences. 1981;**293**:419-476. DOI: 10.1098/rstb.1981.0081. Available from: http://www.jstor.org/stable/2395592

[35] Conley DJ, Paerl HW, Howarth RW, Boesch DF, Seitzinger SP, Havens KE, et al. Controlling eutrophication: Nitrogen and phosphorus. Science. 2009;**323**:1014-1015. DOI: 10.1126/science.1167755

[36] Ahn C-Y, Lee CS, Choi JW, Lee S, Oh H-M. Global occurrence of harmful cyanobacterial blooms and N, P-limitation strategy for bloom control. Korean Journal of Environmental Biology. 2015;**33**:1-6. DOI: 10.11626/KJEB.2015.33.1.001

[37] Tempest D, Neijssel O, Zevenboom W. Properties and performance of microorganisms in laboratory culture; their relevance to growth in natural ecosystems. Symposium of the Society for General Microbiology. 1983;**34**:119-152

[38] Paerl HW. Mitigating harmful cyanobacterial blooms in a human-and climatically-impacted world. Life. 2014;**4**(4):988-1012. DOI: 10.3390/life4040988

[39] Crettaz Minaglia M, Rosso L, Aranda J, Goñi S, Sedan D, Andrinolo D, et al. Mathematical modeling of *Microcystis aeruginosa* growth and [D-Leu1] microcystin-LR production in culture media at different temperatures. Harmful Algae. 2017;**67**:13-25. DOI: 10.1016/j.hal.2017.05.006

[40] Lürling M, Eshetu F, Faassen EJ, Kosten S, Huszar VLM. Comparison of cyanobacterial and green algal growth rates at different temperatures. Freshwater Biology. 2013;**58**(3):552-559. DOI: 10.1111/j.1365-2427.2012.02866.x

[41] Paerl HW, Huisman J. Mini review: Climate change: A catalyst for global expansion of harmful cyanobacterial blooms. Environmental Microbiology Reports. 2009;**1**(1):27-37. DOI: 10.1111/j.1758-2229.2008.00004.x

[42] O'Neil JM, Davis TW, Burford MA, Gobler CJ. The rise of harmful

cyanobacteria blooms: The potential roles of eutrophication and climate change. Harmful Algae. 2012;**14**:313-334. DOI: 10.1016/j.hal.2011.10.027

[43] Gorham PR. Toxic algae. In: Jackson DF, editor. Algae and Man. New York, N.Y: Plenum Press; 1964. pp. 307-336. DOI: 10.1007/978-1-4684-1719-7_15

[44] Runnegar MTC, Falconer IR, Jackson ARB, McInnes A. Toxin production by *Microcystis aeruginosa* cultures. Toxicon. 1983;**21**(Suppl 3):377-380. DOI: 10.1016/0041-0101(83)90233-7

[45] Codd GA, Poon GK. Cyanobacterial toxins. Proceedings of the Phytochemical Society of Europe. 1988;**28**:283-296

[46] Sivonen K. Effects of light, temperature, nitrate orthophosphate, and bacteria on growth and hepatotoxin production by *Oscillatoria agardhii* strains. Applied and Environmental Microbiology. 1990;**56**:2658-2666

[47] Rapala J, Sivonen K, Lyra C, Niemela SI. Variation of microcystins, cyanobacterial hepatoxins, in *Anabaena* spp. as a function of growth stimuli. Applied and Environmental Microbiology. 1997;**63**:2206-2212

[48] Lehman PW, Boyer G, Satchwell M, Waller S. The influence of environmental conditions on the seasonal variation of *Microcystis* cell density and microcystins concentration in San Francisco estuary. Hydrobiologia. 2008;**600**:187-204. DOI: 10.1007/s10750-007-9231-x

[49] Wang Q, Niu Y, Xie P, Chen J, Ma ZM, Tao M, et al. Factors affecting temporal and spatial variations of microcystins in Gonghu Bay of Lake Taihu, with potential risk of microcystin contamination to human health. Scientific World Journal. 2010;**10**:1795-1809. DOI: 10.1100/tsw.2010.172

[50] Giannuzzi L, Krockc B, Crettaz Minaglia MC, Rosso L, Houghton C, Sedan D, et al. Growth, toxin production, active oxygen species and catalase activity of *Microcystis aeruginosa* (Cyanophyceae) exposed to temperature stress. Comparative Biochemistry and Physiology, Part C. 2016;**189**:22-30. DOI: 10.1016/j.cbpc.2016.07.001

[51] van der Westhuizen AJ, Eloff JN, Krüger GHJ. Effect of temperature and light (fluence rate) on the composition of the toxin of the cyanobacterium *Microcystis aeruginosa* (UV-006). Archiv für Hydrobiologie. 1986;**108**(2):145-154. DOI: 10.1007/BF00395897

[52] Mowe M, Porojan C, Abbas F, Mitrovic S, Lim R, Furey A, et al. Rising temperatures may increase growth rates and microcystin production in tropical *Microcystis* species. Harmful Algae. 2015;**50**:88-98. DOI: 10.1016/j.hal.2015.10.011

[53] Tsukada H, Tsujimura S, Nakahara H. Effect of nutrient availability on the C, N, and P elemental ratios in the cyanobacterium *Microcystis aeruginosa*. Limnology. 2006;7(3):185, 17-192. DOI: 10.1007/s10201-006-0188-7

[54] Seviour RJ, Nielsen PH, editors. Microbial Ecology of Activated Sludge. London, UK: IWA Publishing; 2010. DOI: 10.2166/9781780401645

[55] Ghaffar S, Stevenson RJ, Khan Z. Effect of phosphorus stress on *Microcystis aeruginosa* growth and phosphorus uptake. PLoS One. 2017;**12**(3):e0174349. DOI: 10.1371/journal.pone.0174349

[56] Zwietering MH, de Koos JT, Hasenack BE, de Wit JC, van't Riet K. Modelling of the bacterial growth as a function of temperature. Applied and Environmental Microbiology. 1991;**57**:1094-1101

[57] Zwietering MH, Jongenburger FM, Roumbouts M, Van't Riet K. Modelling

of the bacterial growth curve. Applied and Environmental Microbiology. 1990;**56**:1875-1881

[58] Whiting RC. Microbial modeling in foods. Critical Reviews in Food Science and Nutrition. 1995;**35**:467-494. DOI: 10.1080/10408399509527711

[59] Giannuzzi L, Pinotti A, Zaritzky N. Mathematical modelling of microbial growth in packaged refrigerated beef stored at different temperatures. International Journal of Food Microbiology. 1998;**39**:101-110. DOI: 10.1016/s0168-1605(97)00127-x

[60] Fujimoto N, Sudo R, Sugiura N, Inamori Y. Nutrient-limited growth of *Microcystis aeruginosa* and *Phomzidium tenue* and competition under various N:P supply ratios and temperatures. Limnology and Oceanography. 1997;**42**:250-256. DOI: 10.4319/lo.1997.42.2.0250

[61] Lyck S. Simultaneous changes in cell quotas of microcystin, chlorophyll a, protein and carbohydrate during different growth phases of a batch culture experiment with *Microcystis aeruginosa*. Journal of Plankton Research. 2004;**26**(7):727-736. DOI: 10.1093/plankt/fbh071

[62] Jiang Y, Ji B, Wong RNS, Wong NH. Statistical study on the effects of environmental factors on the growth and microcystins production of bloom-forming cyanobacterium—*Microcystis aeruginosa*. Harmful Algae. 2008;**7**:127-136. DOI: 10.1016/j.hal.2007.05.012

[63] Jähnichen S, Long BM, Petzoldt T. Microcystin production by *Microcystis aeruginosa*: Direct regulation by multiple environmental factors. Harmful Algae. 2001;**12**:95-104. DOI: 10.1016/j.hal.2011.09.002

[64] Bortoli S, Oliveira-Silva D, Krügerd T, Dörra FA, Colepicolo P, Volmer D, et al. Growth and microcystin production of a Brazilian *Microcystis aeruginosa* strain (LTPNA 02) under different nutrient conditions. Revista Brasileira de Farmacognosia. 2014;**24**:389-398. DOI: 10.1016/j.bjp.2014.07.019

[65] Canale RP, Vogel AH. Effects of temperature on phytoplankton growth. Journal of the Environmental Engineering Division, American Society of Civil Engineers. 1974;**100**:229-241

[66] Reynolds CS. The Ecology of Freshwater Phytoplankton. Cambridge: Cambridge University Press; 1984. DOI: 10.1046/j.1365-2427.2002.00888.x

[67] Shao J, Li R, Lepo J, Gu D. Review potential for control of harmful cyanobacterial blooms using biologically derived substances: Problems and prospects. Journal of Environmental Management. 2013;**125**:149-155. DOI: 10.1016/j.jenvman.2013.04.001

[68] Anderson DM. Turning back the harmful red tides. Nature. 1997;**38**:513-514. DOI: 10.1038/41415

[69] Jančula D, Marsálek B. Critical review of actually available chemical compounds for prevention and management of cyanobacterial blooms. Chemosphere. 2011;**85**:1415-1422. DOI: 10.1016/j.chemosphere.2011.08.036

[70] Zamyadi A, Dorner S, Sauve S, Ellis D, Bolduc A, Bastien C, et al. Species-dependence of cyanobacteria removal efficiency by different drinking water treatment processes. Water Research. 2013;**47**:2689-2700. DOI: 10.1016/j.watres.2013.02.040

[71] Matthijs HCP, Jančula D, Visser PM, Maršálek B. Existing and emerging cyanocidal compounds: New perspectives for cyanobacterial bloom mitigation. Aquatic Ecology. 2016;**50**:443-460. DOI: 10.1007/s10452-016-9577-0

[72] Shao J, Li R, Lepo J, Gu J. Potential for control of harmful cyanobacterial blooms using biologically derived substances: Problems and prospects. Journal of Environmental Management. 2013;**125**:149-155. DOI: 10.1016/j.jenvman.2013.04.001

[73] Li J, Shimizu K, Maseda H, Lu Z, Utsumi M, Zhang Z, et al. Investigations into the biodegradation of microcystin-lr mediated by the biofilm in wintertime from a biological treatment facility in a drinking-water treatment plant. Bioresource Technology. 2012;**106**:27-35. DOI: 10.1016/j.biortech.2011.11.099

[74] Ame MV, Galanti LN, Menone ML, Gerpe MS, Moreno VJ, Wunderlin DA. Microcystin–LR,–RR,–YR and–LA in water samples and fishes from a shallow lake in Argentina. Harmful Algae. 2010;**9**(1):66-73. DOI: 10.1016/j.hal.2009.08.001

[75] Okano K, Shimizu K, Kawauchi Y, Maseda H, Utsumi M, Zhang Z, et al. Characteristics of a microcystin-degrading bacterium under alkaline environmental conditions. Journal of Toxicology. 2009:1-8. DOI: 10.1155/2009/954291

[76] Ridge I, Barrett PRF. Algal control with barley straw. Aspects of Applied Biology. 1992;**29**:457-462

[77] Schrader KK, de Regt MQ, Tidwell PR, Tucker CS, Duke SO. Selective growth inhibition of the musty-odor producing cyanobacterium *Oscillatoria* cf. chalybea by natural compounds. Bulletin of Environmental Contamination and Toxicology. 1998;**60**:651-658. DOI: 10.1007/s001289900676

[78] Nakai S, Inoue Y, Hosomi M, Murakami A. *Myriophyllum spicatum* released allelopathic polyphenols inhibiting growth of blue-green algae *Microcystis aeruginosa*. Water Research. 2000;**34**:3026-3032. DOI: 10.1016/s0043-1354(00)00039-7

[79] Mulderij G, Smolders Alfons JP, van Donk E. Allelopathic effect of the aquatic macrophyte, *Stratiotes aloides*, on natural phytoplankton. Freshwater Biology. 2006;**51**:554-561. DOI: 10.1111/j.1365-2427.2006.01510.x

[80] Planas D, Sarhan F, Dube L, Godmaire H, Cadieux C. Ecological significance of phenolic compounds of *Myriophyllum spicatum*. Verhandlungen des Internationalen Verein Limnologie. 1981;**21**:1492-1496. DOI: 10.1080/03680770.1980.11897219

[81] Li FM, Hu HY. Isolation and characterization of a novel antialgal allelochemical from *Phragmites communis*. Applied and Environmental Microbiology. 2005;**71**:6545-6553. DOI: 10.1128/aem.71.11.6545-6553.2005

[82] Gross EM, Meyer H, Schilling G. Release and ecological impact of algicidal hydrolysable polyphenols in *Myriophyllum spicatum*. Phytochemistry. 1996;**41**:133-138. DOI: 10.1016/0031-9422(95)00598-6

Cyanobacteria for PHB Bioplastics Production: A Review

Erich Markl, Hannes Grünbichler and Maximilian Lackner

Abstract

Cyanobacteria, or blue-green algae, can be used as host to produce polyhy-droxyalkanoates (PHA), which are promising bioplastic raw materials. The most important material thereof is polyhydroxybutyrate (PHB), which can replace the commodity polymer polypropylene (PP) in many applications, yielding a bio-based, biodegradable alternative solution. The advantage from using cyanobacteria to make PHB over the standard fermentation processes, with sugar or other organic (waste) materials as feedstock, is that the sustainability is better (compare first-generation biofuels with the feed *vs.* fuel debate), with CO_2 being the only carbon source and sunlight being the sole energy source. In this review article, the state of the art of cyanobacterial PHB production and its outlook is discussed. Thirty-seven percent of dry cell weight of PHB could be obtained in 2018, which is getting close to up to 78% of PHB dry cell weight in heterotrophic microorganisms in fermentation reactors. A good potential for cyanobacterial PHB is seen throughout the literature.

Keywords: polyhydroxybutyrate (PHB), bioplastics, EN13432, biodegradability, organic carbon content, microplastics, cyanobacteria

1. Introduction

Bioplastics [1–3] are either biodegradable, e.g., according to the standard EN13432 [4], or at least partly made from renewable raw materials, e.g., according to ASTM D6866 [5]. Although their market share today is only approx. 2%, they see two-digit growth figures [6]. The sustainability of bioplastics is reviewed in [7]. Plastics in general and their composites are a large and important class of materials. The global production volume exceeds 300 million tons/year [8]. For a bioplastics material to have a major impact, it has to match the key properties of one of the commodity plastics such as PP, PE, PVC, PS or PET. This is the case with polyhydroxyalkanoates (PHA), which have the potential to replace mass polymer PP in many applications. Polyhydroxybutyrate (PHB) is the most important representative of PHA.

Cyanobacteria [9–11] are a phylum of bacteria that obtain their energy through pho-tosynthesis, and they are the only photosynthetic prokaryotes that can produce oxygen. The name "cyanobacteria" is derived from the Greek word for "blue," which is the color of cyanobacteria. Cyanobacteria are prokaryotes, and they are also called "blue-green algae," though the term "algae" is not correct technically, as it only includes eukaryotes.

It was discovered that cyanobacteria can produce polyhydroxyalkanoates (PHA) photoautotrophically [12], with the potential for CO_2 recycling and bioplastics pro-duction. This chapter is an up-to-date review on PHB production from cyanobacteria, since the last review article on this topic [13] was written already 5 years ago.

2. PHB, a commodity bioplastics for mass market products?

Today, thermoplastic starch (TPS) and polylactic acid (PLA) are the two domi-
nant biodegradable bioplastics materials. Partly, bio-based PET (see, for example,
the PlantBottle™ project) and "Green PE," a polyethylene made from sugarcane-
derived ethanol in Brazil, are the two most common nondegradable, but bio-based
plastics. PHB has striking similarities to PP and has therefore been envisaged as
potential replacement candidate for PP by Markl et al. [14], for instance, in biomedical, agricultural, and industrial applications [15]. The following **Table 1** shows
a comparison of PHB and PP.

Property	PHB	PP
Crystalline melting point (°C)	175	176
Crystallinity (%)	80	70
Molecular weight (Daltons)	5×10^5	2×10^5
Glass transition temperature (°C)	4	−10
Density (g/cm³)	1.250	0.905
Flexural modulus (GPa)	4.0	1.7
Tensile strength (MPa)	40	38
Extension to break (%)	6	400
Ultraviolet resistance	good	poor
Solvent resistance	poor	good

Table 1.
Properties of PHB compared to those of PP (source: [16]).

Properties	PHB	PHBV	PHB4B	PHBHx	PP
Melting temperature (°C)	177	145	150	127	176
Glass transition temperature (°C)	2	−1	−7	−1	−10
Crystallinity (%)	60	56	45	34	50–70
Tensile strength (MPa)	43	20	26	21	38
Extension to break (%)	5	50	444	400	400

Table 2.
Property modification by copolymerization (source: [13]).

The low elongation and break and the brittleness of PHB are limitations. These,
however, can be overcome by using other PHA, blends of copolymers, see **Table 2**.
Apart from short-chain-length PHA, there are medium- and long-chain-
length variants, too, [17], so that material properties can be tailored in a wide
spectrum.
The majority of PP is used in short-lived plastic products such as rigid packaging, which partly end up in nature. A biodegradable alternative can be a sensible
material solution. Since PHA can be selected and customized for various applications, and also blended, co-polymerized and compounded, it is estimated that up to
90% of all PP applications can be covered by PHA and to a large extent thereof by
PHB. A disadvantage of PHB is its high production cost. In [15], ways to make PHA
production more cost-competitive are listed (see **Table 3**).

Technology	Reasons and/or purpose	Methodology
High cell density fermentation	Achieve effective growth and cells recovery	Manipulation on quorum sensing and cell oxygen uptake mechanisms
Growth cells in low cost substrates or mixed substrates	Substrates contributed to over 60% of PHA cost	Screening targeted substrates utilizing bacteria able to produce high content PHA
Fast growing cells	Reduce fermentation duration and avoid microbial contamination	Minimizing bacterial genome, changing cell growth patterns
Fast growing CO_2 utilizing bacteria able to produce PHA	CO_2 is a free substrate	Manipulating the CO_2 uptake mechanism such as carboxysomes, etc.
Open (unsterile) and continuous fermentation process	To save sterilization energy, reduce fermentation complexity and improve process effectiveness	Screening for PHA producers able to grow fast in extreme environments such as high or low pH and temperature, high osmotic pressure, etc.
PHA synthesis induced by oxygen limitation	Oxygen is a limited factor in all high cell density growth	Place PHA synthesis operons behind microaerobic promotor
Ultrahigh PHA accumulation (over 95% PHA in cell dry weight)	To avoid expensive and complicated downstream PHA purification process	Manipulating the PHA synthesis mechanism and PHA synthases
Increase substrate (mostly carbon sources) to PHA conversion efficiency	Substrates contributed to over 60% of PHA cost	Removing pathways that consume substrates for non-PHA metabolisms, and/or reinforce PHA synthesis flux
Enlarging the PHA production cells	To allow more cellular space for PHA accumulation, this also allows easy cells recovery	Engineering the cell division patterns and/or cytoskeletons
Inducible cell flocculation	Allow easy biomass recovery after fermentation	Inducible expression of surface displaying adhesive proteins
Inducible cell lysis	Allow easy PHA granules recovery after biomass harvest	Inducible expression of cell lysis proteins
Cell disruption by PHA hyperproduction	Save the biomass harvest process	Manipulating the PHA synthesis mechanism and PHA synthases
Extracellular PHA production	Not limited by a small cellular space, also for easy PHA granule recovery	Need new PHA synthesis mechanisms
Large PHA granules	Allow easy recovery of PHA granules from lysis broth	Manipulating the formation of PHA granules associated proteins
A synthetic cell combining the above properties	Achieve up-stream and down-stream competitivenesses	An artificial cell with assembled functional DNA

Table 3.
Technology to be developed to lower PHA production cost (reproduced with permission from [15]).

Avoiding feedstock costs and using CO_2 as sole carbon source are described as strong potential here.

In general, organic carbon feedstocks can yield high PHB contents in microorganisms. For instance, Bhati et al. produced 78% PHB of dry well weight with *Nostoc muscorum* Agardh [18].

An alternative production pathway for PHB is a catalytic one [19, 20]. Both the fermentation and the catalytic process yield an expensive PHB product, which is hard to sell as it competes with low-price commodities such as PE and PP for packaging applications, which are very cost-sensitive.

3. PHB production by cyanobacteria: current state of knowledge

It is known that cyanobacteria can produce PHB as an intracellular energy and carbon storage compound [21] (see **Figure 1**).

Reference [23] discusses the use of cyanobacteria to produce chemicals. Cyanobacteria show several industrially relevant benefits compared to their plant counterparts, including a faster growth rate, higher CO_2 utilization and greater amenability to genetic engineering [24, 25].

Table 4 shows compounds that can be produced by cyanobacteria photoautotrophically [26].

In 2013, a review on the production of poly-β-hydroxybutyrates from cyanobacteria for the production of bioplastics was published [13]. Meanwhile, significant improvements have been implemented.

In 2018, Troschl et al. could report 12.5% PHB cry well weight [21]. In the same year, Kamravamanesh et al. have shown that the cyanobacterium *Synechocystis* sp. PCC 6714 can produce up to 37% dry cell weight of PHB with CO_2 as the only carbon source [27, 28], which is significantly above the other reported values from literature. The strain had been subjected to UV light mutations to increase the PHB productivity. Prior to that work, the thermophilic cyanobacterium,

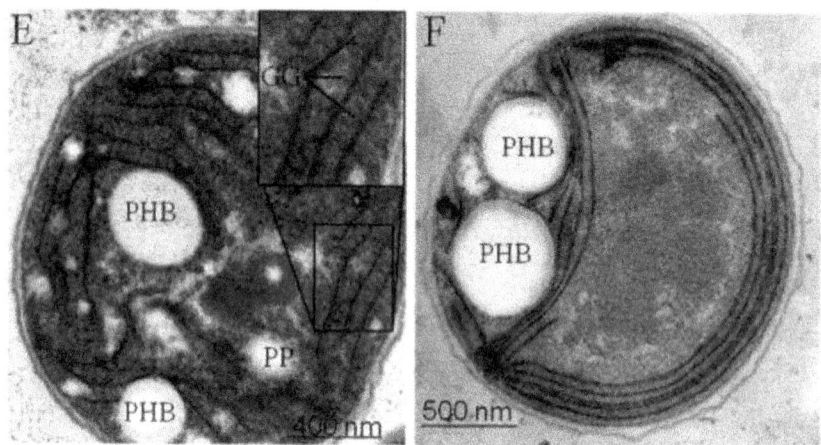

Figure 1.
PHB granules in cyanobacteria. Left: Wild type. Right: Mutant (reproduced with permission from [22]).

Strain	Compound	Titer (g/L)
2973	sucrose	3.3
6803	3-hydroxypropionic acid	0.8
	ethanol	5.5
	isobutanol	0.6
	lactic acid	0.8
	limonene	0.007
7002	2,3-butanediol	1.6
	alpha bisabolene	0.0006
	fatty acids	0.1
	glycogen	1.8
		3.0
		3.5
	limonene	0.004
	lysine	0.4
	mannitol	1.1
	poly-3-hydroxybutyrate	0.05
7942	2,3-butanediol	3.0[a]
		5.7[b]
		12.6[c]
	alpha-farnesene	0.005
	ethanol	0.07
	fatty acid ethyl esters	0.01
	isobutyraldehyde	1.1
	isoprene	1.3
	limonene	0.005
	squalene	0.05
	succinate	0.4
		7.3
	sucrose	2.6
		0.8

Table 4.
Compounds that could be produced by cyanobacteria (reproduced with permission from [26]).

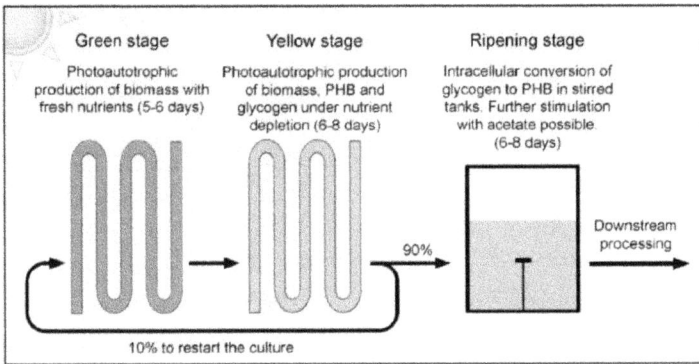

Figure 2.
Operation mode for PHB production from cyanobacteria. The ripening tank is used for PHB production at a later stage, where no CO_2 is consumed, but glycogen gets converted into PHB (reproduced with permission from [18]).

Synechococcus sp. MA19, was reported to have achieved 27% of dry cell weight PHB [29]. It was reported that, originally, the MA 19 was isolated from a hot spring in Japan (Miyakejima). However, neither the authors of this paper nor other researchers [30] were able to obtain a sample from that strain in 2016–2018, despite high efforts, so currently, Kamravamanesh's strain *Synechocystis* sp. PCC 6714 can be considered the cyanobacterium with the highest PHB content. A high PHB content is advantageous for downstream processing in terms of energy efficiency, for instance, or product quality.

Genetic engineering is commonly deployed to increase the yield of PHB compared to wild types [26, 31, 32]. Also, bioprocess optimization is carried out [27, 28]. Growth is typically followed by nitrogen and/or phosphorous limitation. Also, "feast and famine" strategies concerning the carbon source are applied [33].

Reference [34] discusses the use of consortia of cyanobacteria and heterotrophic bacteria for stable PHB production.

The modeling of cyanobacterial PHB production is discussed in [35].

A possible growth system for PHB from cyanobacteria is presented in [18], see **Figure 2** below.

The study in [18] uses long-term, non-sterile cultivation of *Synechocystis* sp. CCALA192 in a tubular photobioreactor for PHB production. Another concept would be open pond photobioreactors like open pond raceways. Different photobioreactor setups are reviewed in [18, 36–39]. A promising alternative is an integrated algae-based biorefinery, e.g., for the production of biodiesel, astaxanthin and PHB as presented by [40] or [41].

4. PHB production by cyanobacteria: an outlook

A major unsolved issue is the downstream processing of the cyanobacteria, i.e., how to get the bioplastics material out of the cyanobacteria (see **Figure 3**).

In Ref. [23], photomixotrophic conditions to increase cyanobacterial production rate and yield are reviewed. Supplementation with fixed carbon sources gives additional carbon building blocks and energy to speed up production. Photomixotrophic production was found to increase titers up to fivefold over traditional autotrophic conditions [23], so there is a strong future potential in this mode for cyanobacteria.

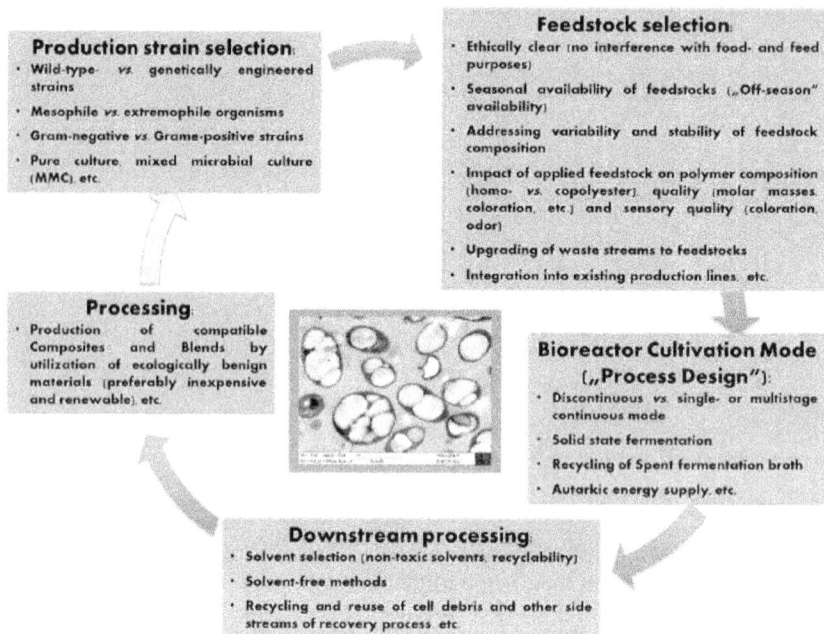

Figure 3.
Schematic illustration of factors impacting sustainability of PHA production (reproduced with permission from [42]).

5. Conclusions

This chapter has presented an update on PHB production by cyanobacteria, a process route which can be more sustainable than catalytic production from CO or fermentation from sugar compounds. It is expected that PHB and its compounds will gradually replace PP in many large volume applications. Genetic engineering can increase the yield of PHB in cyanobacteria; however, the downside is that approval for large-scale cultivation in (cost- and energy-efficient) open growth systems will be difficult to obtain in most countries, so technologies avoiding genetic engineering seem to be most promising for commercial development.

Acknowledgements

Financial support from Wirtschaftsagentur Wien is gratefully acknowledged.

Conflict of interest

The authors declare that they have no conflict of interest.

Author details

Erich Markl*, Hannes Grünbichler and Maximilian Lackner
University of Applied Sciences Technikum Wien, Hoechstaedtplatz, Vienna, Austria

*Address all correspondence to: erich.markl@technikum-wien.at

IntechOpen

References

[1] Lackner M. Bioplastics—Biobased plastics as renewable and/or biodegradable alternatives to petroplastics. In: Kirk-Othmer Encyclopedia of Chemical Technology. New York, USA: Wiley; 2015

[2] Ashter SA. Introduction to Bioplastics Engineering. Oxford, UK: William Andrew; 2016. ISBN: 978-0323393966

[3] Kabasci S, editor. Bio-Based Plastics: Materials and Applications. Weinheim, Germany: Wiley VCH; 2013. ISBN: 9781119994008

[4] DIN EN 13432-2000. Packaging—Requirements for Packaging Recoverable Through Composting and Biodegradation—Test Scheme and Evaluation Criteria for the Final Acceptance of Packaging; German Version EN 13432:2000. https://www.beuth.de/de/norm/din-en-13432/32115376 Accessed: August 1, 2018

[5] ASTM D6866-18. Standard Test Methods for Determining the Biobased Content of Solid, Liquid, and Gaseous Samples Using Radiocarbon Analysis. https://www.astm.org/Standards/D6866.htm Accessed: August 1, 2018

[6] Endres H-J. Engineering Biopolymers: Markets, Manufacturing, Properties and Applications. Munich, Germany: Hanser Pubn; 2010. ISBN: 978-1569904619

[7] Thakur S, Chaudhary J, Sharma B, Verma A, Thakur VK. Sustainability of bioplastics: Opportunities and challenges. Current Opinion in Green and Sustainable Chemistry. 2018;**13**:68-75

[8] Plastics Europe, The Facts 2017. https://www.plasticseurope.org/application/files/5715/1717/4180/Plastics_the_facts_2017_FINAL_for_website_one_page.pdf

[9] Sharma NK, Rai AK, Stal LJ. Cyanobacteria: An Economic Perspective. Malden, USA: Wiley-Blackwell; 2014. ISBN: 978-1119941279

[10] Nienaber MA, Steinitz-Kannan M. A Guide to Cyanobacteria: Identification and Impact Kindle Edition. Lexington, USA: University Press of Kentucky; 2018. ISBN: 978-0813175591

[11] Tiwari A. Cyanobacteria Nature, Potentials and Applications. Lexington, USA: Astral; 2014. ISBN-13: 978-8170359128

[12] Asada Y, Miyake M, Miyake J, Kurane R, Tokiwa Y. Photosynthetic accumulation of poly-(hydroxybutyrate) by cyanobacteria—The metabolism and potential for CO_2 recycling. International Journal of Biological Macromolecules. 1999;**25**(1-3):37-42

[13] Balaji S, Gopi K, Muthuvelan B. A review on production of poly β hydroxybutyrates from cyanobacteria for the production of bio plastics. Algal Research. 2013;**2**(3):278-285

[14] Markl E, Grünbichler H, Lackner M. PHB—Biobased and biodegradable replacement for PP: A review. Novel Techniques in Nutrition and Food Science (NTNF). 2018;**2**(4):1-4

[15] Możejko-Ciesielska J, Kiewisz R. Bacterial polyhydroxyalkanoates: Still fabulous? Microbiological Research. 2016;**192**:271-282

[16] Kaplan DL. Biopolymers from Renewable Resources. New York, USA: Springer; 1998. ISBN: 978-3540635673

[17] Singh AK, Mallick N. Enhanced production of SCL-LCL-PHA co-polymer by sludge-isolated

Pseudomonas aeruginosa MTCC 7925. Letters in Applied Microbiology. 2008;**46**(3):350-357. DOI: 10.1111/j.1472-765X.2008.02323.x

[18] Bhati R, Mallick N. Poly(3-hydroxybutyrate-co-3-hydroxyvalerate) copolymer production by the diazotrophic cyanobacterium *Nostoc muscorum* Agardh: Process optimization and polymer characterization. Algal Research. 2015;7:78-85. DOI: 10.1016/j.algal.2014.12.003

[19] Wang Y, Yin J, Chen G-Q. Polyhydroxyalkanoates, challenges and opportunities. Current Opinion in Biotechnology. 2014;**30**:59-65

[20] Reichardt R, Rieger B. Poly (3-Hydroxybutyrate) from carbon monoxide. In: Rieger B, Künkel A, Coates GW, Reichardt R, Dinjus E, Zevaco TA, editors. Synthetic Biodegradable Polymers. New York, USA: Springer; 2012 https://link.springer.com/chapter/10.1007/12_2011_127. ISBN 978-3-642-27154-0

[21] Troschl C, Meixner K, Fritz I, Leitner K, Drosg B. Pilot-scale production of poly-β-hydroxybutyrate with the cyanobacterium *Synechocystis* sp. CCALA192 in a non-sterile tubular photobioreactor. Algal Research. 2018;**34**:116-125

[22] Damrow R, Maldener I, Zilliges Y. The multiple functions of common microbial carbon polymers, glycogen and PHB, during stress responses in the non-diazotrophic cyanobacterium *Synechocystis* sp. PCC 6803. Frontiers in Microbiology. 2016;7:966. DOI: 10.3389/fmicb.2016.00966

[23] Matson MM, Atsumi S. Photomixotrophic chemical production in cyanobacteria. Current Opinion in Biotechnology. 2018;**50**:65-71

[24] Field CB, Behrenfeld MJ, Randerson JT, Falkowski P. Primary production of the biosphere: Integrating terrestrial and oceanic components. Science. 1998;**281**:237-240

[25] Heidorn T, Camsund D, Huang HH, Lindberg P, Oliveira P, Stensjo K, et al. Synthetic biology in cyanobacteria engineering and analyzing novel functions. Methods in Enzymology. 2011;**497**:539-579

[26] Carroll AL, Case AE, Zhang A, Atsumi S. Metabolic engineering tools in model cyanobacteria. Metabolic Engineering. In press, corrected proof, Available online 26 March 2018

[27] Kamravamanesh D, Pflügl S, Nischkauer W, Limbeck A, Lackner M, Herwig C. Photosynthetic poly-β-hydroxybutyrate accumulation in unicellular cyanobacterium *Synechocystis* sp. PCC 6714. AMB Express. 2017;7(1):143. DOI: 10.1186/s13568-017-0443-9. Epub 2017 Jul 6

[28] Kamravamanesh D, Pflügl S, Kovacs T, Druzhinina I, Kroll P, Maximilian L, et al. Increased poly-beta-hydroxybutyrate production from CO_2 in randomly mutated cells of cyanobacterial strain *Synechocystis* sp. PCC 6714: Mutant generation and characterization. Bioresource Technology. 2018;**266**:34-44. DOI: 10.1016/j.biortech.2018.06.057

[29] Miyake M, Erata M, Asada Y. A thermophilic cyanobacterium, Synechococcus sp. MA19, capable of accumulating poly-β-hydroxybutyrate. Journal of Fermentation and Bioengineering. 1996;**82**(5):512-514

[30] Fritz I. Universität für Bodenkultur Wien. BOKU, University of Natural Resources and Life Sciences, Vienna, Private Communication. 2018

[31] Hondo S, Takahashi M, Osanai T, Matsuda M, Asayama M. Genetic engineering and metabolite profiling for overproduction of polyhydroxybutyrate

in cyanobacteria. Journal of Bioscience and Bioengineering. 2015;**120**(5):510-517

[32] Angermayr SA, Gorchs Rovira A, Hellingwerf KJ. Metabolic engineering of cyanobacteria for the synthesis of commodity products. Trends in Biotechnology. 2015;**33**(6):352-361

[33] Arias DM, Fradinho JC, Uggetti E, García J, Reis MAM. Polymer accumulation in mixed cyanobacterial cultures selected under the feast and famine strategy. Algal Research. 2018;**33**:99-108

[34] Weiss TL, Young EJ, Ducat DC. A synthetic, light-driven consortium of cyanobacteria and heterotrophic bacteria enables stable polyhydroxybutyrate production. Metabolic Engineering. 2017;**44**:236-245

[35] Carpine R, Raganati F, Olivieri G, Hellingwerf KJ, Marzocchella A. Poly-β-hydroxybutyrate (PHB) production by *Synechocystis* PCC6803 from CO₂: Model development. Algal Research. 2018;**29**:49-60

[36] Diehl S, Zambrano J, Carlsson B. Analysis of photobioreactors in series. Mathematical Biosciences. In press, accepted manuscript, Available online 27 July 2018

[37] Wolf J, Stephens E, Steinbusch S, Yarnold J, Hankamer B. Multifactorial comparison of photobioreactor geometries in parallel microalgae cultivations. Algal Research. 2016;**15**:187-201

[38] Perez-Castro A, Sanchez-Moreno J, Castilla M. PhotoBioLib: A Modelica library for modeling and simulation of large-scale photobioreactors. Computers & Chemical Engineering. 2017;**98**:12-20

[39] Acién FG, Molina E, Reis A, Torzillo G, Masojídek J. Photobioreactors for the production of microalgae.

In: Microalgae-Based Biofuels and Bioproducts. Duxford, UK: Woodhead Publishing; 2017. pp. 1-44

[40] García Prieto CV, Ramos FD, Estrada V, Villar MA, Soledad Diaz M. Optimization of an integrated algae-based biorefinery for the production of biodiesel, astaxanthin and PHB. Energy. 2017;**139**(15):1159-1172

[41] Meixner K, Kovalcik A, Sykacek E, Gruber-Brunhumer M, Drosg B. Cyanobacteria biorefinery—Production of poly(3-hydroxybutyrate) with *Synechocystis* salina and utilisation of residual biomass. Journal of Biotechnology. 2018;**265**:46-53

[42] Koller M, Maršálek L, de Sousa Dias MM, Braunegg G. Producing microbial polyhydroxyalkanoate (PHA) biopolyesters in a sustainable manner. New Biotechnology. 2017;**37**(Part A): 24-38. DOI: 10.1016/j.nbt.2016.05.001

Chapter 4

CO$_2$ Capture for Industries by Algae

Vetrivel Anguselvi, Reginald Ebhin Masto, Ashis Mukherjee and Pradeep Kumar Singh

Abstract

The increased usage of fossil fuels has led to increase in the concentration of CO$_2$, which is a greenhouse gas responsible for global warming. Algae-based CO$_2$ conversion is a cost-effective option for reducing carbon footprint. In addition, algae-based CO$_2$ mitigation strategy has the potential to obtain valuable products at the end of the process. In the present study, freshwater algal species were isolated and identified for CO$_2$ capture, such as *Hydrodictyon*, *Spirogyra*, *Oscillatoria*, *Oedogonium*, and *Chlorella*. The algal strains were screened based on different parameters like fast growth rate, high rate of photosynthesis, strong tolerance to the trace constituents of other gases (gaseous hydrocarbons, NOx, SOx, etc.), high temperature tolerance, and possibility to produce high value products, etc. The study involves integrated methods for utilizing 90–99% CO$_2$ from a natural gas processing industry (GAIL India, Ltd.) as well as 13–15% of CO$_2$ from flue gas of thermal power plants (*Chandrapura* and Santaldih *Thermal Power Station*) as carbon nutrient source along with the additional nutritional supplements. A 400-ml and 25-l flat panel photo-bioreactor (PSI Photo-bioreactors) was used for CO$_2$ capture. After CO$_2$ capture, the algal biomass was used to extract value-added products such as amino acid rich feed, algal oil, algal pellets, etc.

Keywords: algae, CO$_2$, flue gas, capture, petrochemical industries, thermal power station

1. Introduction

Greenhouse gas emissions by industries and human activities make the planet warmer. CO$_2$ emitting industries are contributing a major role in the increase of greenhouse gases in the atmosphere for several decades. The largest source of greenhouse gas emissions from human activities in the world is burning fossil fuels for electricity, heat, transportation, and domestic uses. Carbon sequestration, capturing, and storing carbon emitted from the global energy system could be a major tool for reducing atmospheric CO$_2$ concentration. The conventional CO$_2$ sequestration processes like geological sequestration are highly power intensive and therefore expensive. While chemical and physical means exist to capture CO$_2$ from smoke stack emissions, the cost of utilizing these technologies would result in a significant increase in the cost of power. The need for CO$_2$ management, in particular capture and storage, is currently an important technological, economical, and global issue and will continue to be so until alternative energy sources diminish the need for fossil fuels. As microalgae grow in aqueous environments, directly passing CO_2-rich gases through this medium is a very efficient way of capturing the CO_2 in those streams.

Algae-based carbon di-oxide (CO_2) sequestration has gained more interest due to its capability to utilize CO_2 as carbon source, higher photosynthetic efficiency, high CO_2 fixation capacities and optimal culture condition, higher growth rates than conventional crop plants and biomass produced can be used as a feedstock for other value added products such as biofuel and chemicals [1]. In recent years, cultivation of microalgae has received renewed attention on account of its possibility as a feasible CO_2 sequestration technology. Under phototrophic growth conditions, microalgae absorb solar energy, and assimilate CO_2 from air and nutrients from aquatic habitats. One kilogram of algal dry cell weight utilizes around 1.83 kg of CO_2. As per the available literature, an area of 1 Acre (4000 m^2) shall be able to capture about 2.7 tons/day of CO_2 [2]. Algae are receiving wide attention as a source of biomass protein for use in animal feeds and foods [3, 4]. In addition, algae-based CO_2 mitigation strategy has the potential to obtain valuable products from the algal biomass that can be used to generate revenues, thereby making this route feasible. This route could also provide solutions to another major problem viz. high dependency on fossil fuels.

As microalgae grow in aqueous environments, directly passing CO_2-rich gases through this medium is a very efficient way of capturing the CO_2 in those streams. Algae-based CO_2 conversion offers a cost-effective option toward reducing our carbon footprint. In addition, algae-based CO_2 mitigation strategy has the potential to obtain valuable products at the end of the process. Thus, any value addition achieved through such route could also provide solutions to the other major problem viz. high dependency on fossil fuels. Microalgae utilize CO_2 for photosynthesis. Algae-based CO_2 capture is one of the viable options for anthropogenic CO_2 conversion. While microalgal cultivation is expensive, microalgae biomass can be utilized to produce a variety of high value commercial products (algal fuel, protein-rich algal food, animal feed, algae-based medicines, etc.) that can be used to generate revenues, thereby making this route feasible.

At present, more number of researches work is in progress globally to develop and commercialize algae-based carbon conversion technology. Large companies of United States such as Duke Energy, the third largest electric power holding company, are working on various aspects of carbon conversion technology. Government policies, especially in the US and Europe, are also supporting the growth of algae-based carbon conversion technologies. If the challenges associated with algae-based carbon capture technology are addressed successfully, the technology will present a solution not only to the global warming problem, but also to reduce the fossil fuel demand.

2. Experimental study on CO_2 capture for petrochemical industry and thermal power plants

2.1 Isolation, identification, and selection of the high CO_2-tolerant algal strains

In the present study, freshwater algal species were isolated and identified for CO_2 capture, such as *Hydrodictyon*, *Spirogyra*, *Oscillatoria*, *Oedogonium* and *Chlorella* from freshwater taken from pond in the coal mining area of Dhanbad, Jharkhand, India. The blue-green microalgae, *Oscillatoria*, were used for CO_2 capture study. The stock culture and inoculum were grown in BG11 medium [3] in the required condition. The inoculum was pre-cultured aseptically in 500-ml Erlenmeyer flasks with 200 ml of

12

10

8

n

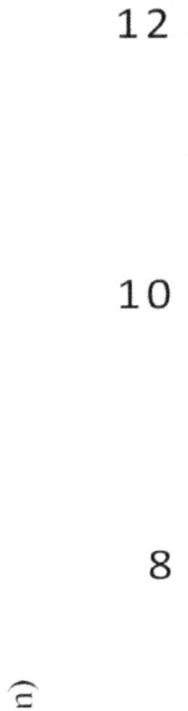

Figure 1.
Scanning electron micrograph of the filament of Oscillatoria.

BG11 medium. The flasks were placed in a 28°C illuminated incubator for 7 days under a 12-h light/12-h dark photoperiod and a light density of 40 $\mu E\ m^{-2}\ s^{-1}$. The BG-11 medium was used for algal culture with the composition of (g/L): $K_2HPO_4.3H_2O$ 0.04, $NaNO_3$ 1.5, $CaCl_2.2H_2O$ 0.036, $MgSO_4.7H_2O$ 0.075, ferric ammonium citrate 0.006, Na_2EDTA 0.001, citric acid 0.006, Na_2CO_3 0.02, pH 7.0. B The identification was done on the basis of the morphological characters of the algae (**Figure 1**).

2.2 Optimization of process parameters for CO₂ conversion

After pre-cultivation, the algal inoculum reached exponential growth phase. One gram of the algal inoculum was collected using centrifugation (4000×g, 4°C, 15 min). The collected algal cells were washed twice with sterile distilled water, and then inoculated into the growth medium. Microalgae growth and composition are affected by several process parameters. Different process parameters were optimized such as pH, light and temperature, nutrient media, culture condition, inoculum volume, inoculum concentration, etc.

2.3 CO₂ capture in photo-bioreactor

Exhaust gas from natural gas processing industry and thermal power plants were collected and analyzed for its composition (**Tables 1** and **2**) (Gas analyzer and GC). Selected species of microalgae were inoculated in a bioreactor and studied for CO_2 capture. Further, a novel "fibrous matrix photo-bioreactor" designed and used for algal culture and CO_2 capture. Novel bioreactor designed and used in the process is having in-built organic fibrous matrix for support of the filamentous algae growth, rapid proliferation, and easy recovery (**Figure 2**).

S. no	Test parameter	Unit	Result
1	Nitrogen	Mole %	0.20
2	Methane	Mole %	0.71
3	Ethane	Mole %	0.20
4	Hydrogen sulfide	ppm	60
5	Amine content	ppm	ND
6	Moisture content	Mole %	1.50
7	Carbon dioxide	Mole %	97.384

Table 1.
Composition of exhaust gas from natural gas processing industry, GAIL (PATA).

Gas	Santaldih TPS	Chandrapura TPS
	Amount (%)	
CO_2	11.93	10–12
O_2	3.96	4–6
SO_2	0.99	Trace
NO_2	0.99	Trace
N_2	82.13	85

Table 2.
Composition of exhaust gas from thermal power plants.

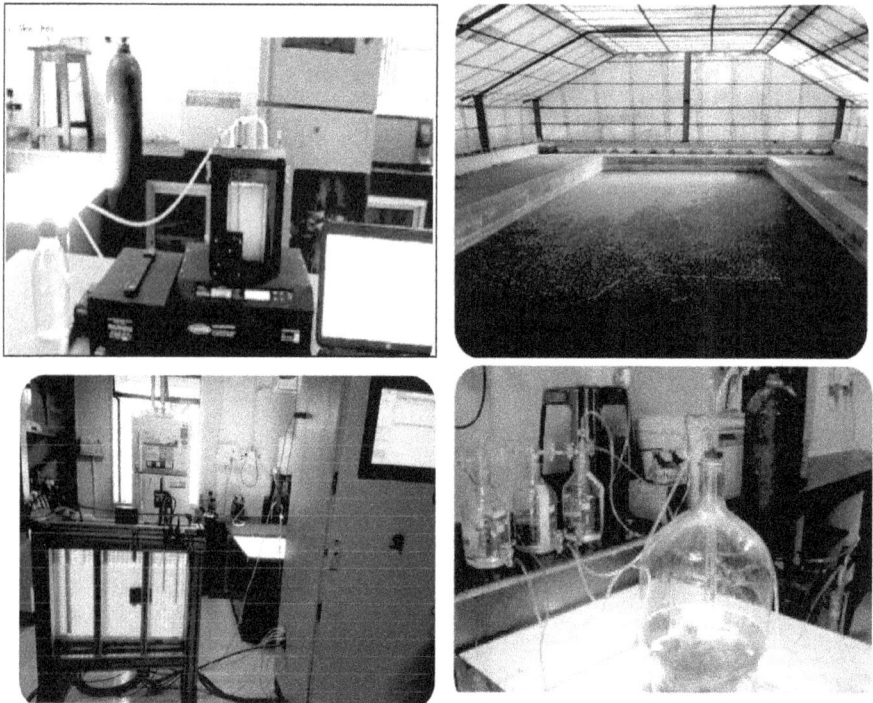

Figure 2.
Light microscope photo of a filament of Oscillatoria.

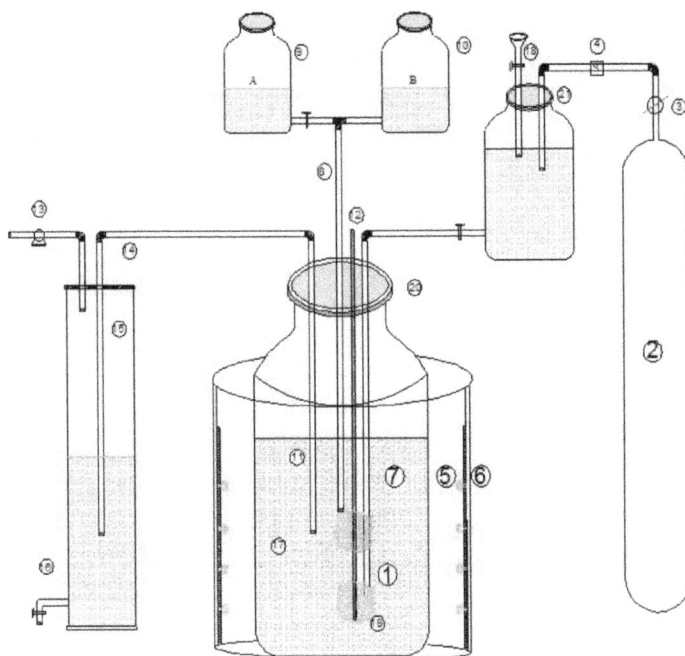

Figure 3.
CO₂ capture and algal fuel production from lab to pilot scale: Facilities at CSIR-CIMFR.

Pigment estimation, protein content, and fatty acid determination were carried out at 5 days interval to determine the growth and subsequently to compare the tolerance of the algal species. The experiment was carried out in flat panel photo-bioreactor (**Figure 3**), which contained 15 l of algal media. Injection of CO_2 rich gas was carried out at a rate of 500 ml/30 min continuously for 48 h. Determination of pH was undertaken at 3 h interval and pigment estimation was carried out at 24, 36, and 48 h intervals.

3. Results and discussion

Oscillatoria is a filamentous blue-green algae with uniserially arranged cells that are not constricted at the cross walls (**Figure 1**). The straight, unbranched, filaments are dark blue-green, covered with a thin hyaline sheath and is not attenuated or capitated at the end. This unbranched filamentous alga occurs singly or in tangled mats. Terminal cells are hemispherical with a slightly thickened membrane on the outer cell envelope (**Figure 4**, Geitler [5]). *Oscillatoria* is common in freshwater environments, including hot springs. It has more than 100 species.

The growth pattern showed that the lag phase of *Oscillatoria* was from 0 to 24 h, exponential phase from 24 to 36 h, the stationary phase from 36 to 60 h and then the decline phase (**Figure 5**).

Growth of algae, in general, depends upon the availability of nitrogen and phosphate (**Table 3**). The increased nutrients result in higher growth and higher biomass of cyanobacteria. Phosphorus was reported to be an essential element for pigment development. Algae are known to assimilate phosphorus in excess of their requirements. The algae (*Oscillatoria*) could not survive in water to which urea was added. Sodium and potassium containing salts along with orthophosphate

Figure 4.
CO_2 capture system using filamentous cyanobacteria: 1—gas absorbed media inlet, 2—gaseous source for feed, 3—gas outlet valve, 4—gas regulator, 5—illumination source, 6—photo-reactor outer wall, 7—growth media, 8—pH control inlet tube, 9—acid source - pH control, 10—alkali source - pH control, 12—media outlet tube, 13—spent media container, 14—spent media exhaust, 15—media inlet tube, 16—media container, 17—media inlet valve, 18—photo-bioreactor vessel, 19—media exhaust tube, 20—fibrous matrix attachment, 21—photo-bioreactor lid, and 22—gas absorption container lid.

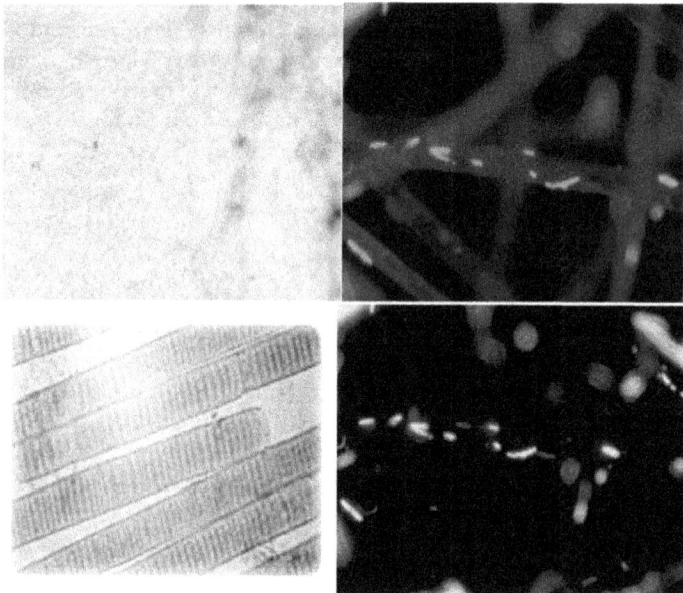

Figure 5.
Growth curve of Oscillatoria.

proved to be a good nutrient for the growth of *Oscillatoria*. Luxuriant growth of *Oscillatoria* was observed in KNO_3 + sodium orthophosphate and $NaNO_3$ + sodium orthophosphate.

Nutrient	Day 0	Day 10	Day 20	Day 30
Control	+	+	+	+
KNO_3 + $NaHPO_4$	+	+++	+++	+++
$NaNO_3$ + $NaHPO_4$	+	+++	+++	+++
Urea	+	+	+	+
Glucose	+	++	++	++

+++: luxuriant growth; ++: moderate growth; +: mild growth.

Table 3.
Effect of nutrients on growth of Oscillatoria.

Metal ion	Days						
	0	5	10	15	20	25	30
Control	+	+	+	+	+	+	+
Ba^+	+	-	+	+	+	+	++
Ca^{++}	+	+	++	++	++	++	++
Co^{++}	+	-	+	+	+	+	++
Cr^{+++}	+	+	+	+	+	+	++
Cu^{++}	+	+	+	+	+	+	++
Fe^{+++}	+	+	++	++	++	++	++
Fe^{++}	+	-	-	+	+	+	++
Mn^{++}	+	-	-	+	+	++	+++
Mg^{++}	+	+++	+++	+	+++	+++	+++
Ni^{++}	+	+	+	+	+	+	++
K^+	+	++	++	+++	+++	+++	+++
Na^+	+	++	++	+++	+++	+++	+++
Sn^{++}	+	+	+	++	++	++	++
Zn^{++}	+	+	+	+	+	+	++
Ag^+	+	-	-	-	-	-	-
Hg^+	+	-	-	+	+	+	+

+++: luxuriant growth; ++: moderate growth; +: mild growth; -: no growth.

Table 4.
Effect of metal ions on growth of Oscillatoria.

3.1 Effect of metal ions on growth of *Oscillatoria*

Maximum growth was observed in salts containing Na^+, Mg^{++}, Ni^{++}, Co^{++}. A drastic increase in chlorophyll content was observed in salt containing Mg^{++}, Mn^{+2} ions. Growth was observed in 15 metal salts out of 16. The algae (*Oscillatoria*) died in salt containing Ag^+. Since, nearly all algae possess chlorophyll and all are expected to carry out molecular phosphate transfers, magnesium is without doubt needed universally by algal species. Net synthesis of RNA may stop immediately following magnesium withdrawal. Potassium is a major cell electrolyte, used for balancing the ionic charge. Sodium plays a key role in maintaining turgor pressure within the organism (**Table 4**).

Vitamin	Day 0	Day 5	Day 10	Day 15
Control	+	+	+	+
Vitamin B12	+	+++	++	+
Vitamin B1	+	+++	+++	+++
Vitamin C	+	+++	+	+

+++: luxuriant growth; ++: moderate growth; +: mild growth.

Table 5.
Growth of Oscillatoria in different vitamins.

Hormones	Day 0	Day 10	Day 20	Day 30
Control	+	+	+	+
IAA*	+	+	++	++
GA**	+	+	++	++

++: moderate growth; +: mild growth. *IAA: Indole-3-acetic acid; **GA: Gibberellic acid.

Table 6.
Growth of Oscillatoria in different hormones.

Surfactants	Day 0	Day 5	Day 10	Day 15
Control	+	+	+	+
Triton X	+	+++	+++	+++
Tween 20	+	+++	+++	+++
Tween 80	+	+++	+++	+++
SDS	+	+++	++	+

+++: luxuriant growth; ++: moderate growth; +: mild growth.

Table 7.
Growth of Oscillatoria in different surfactants.

3.2 Effect of vitamins, growth hormones and surfactants on growth of *Oscillatoria*

Growth was observed in all the three vitamins (B_1, B_{12}, and C). An increase in the chlorophyll content was observed in all the three vitamins till the fifth day. Afterward, luxuriant growth was observed in Vitamin B_1, good growth was observed in Vitamin B_{12}, and moderate in Vitamin C. Vitamin B_{12} acts as a growth factor for all algae as well as higher plants. A good growth was observed in Gibberellic acid as well as Indole-3-acetic acid. Gibberellic acid is a plant growth hormone that influences various developmental processes; while, cyanobacteria respond to IAA in a manner analogous to higher plants. Luxuriant growth was observed in all the four surfactants with increase in the pigment content of *Oscillatoria*. Almost equal growth was observed (visually) up to 7 days. At the tenth day, moderate growth was observed in SDS while luxuriant growth was observed in Triton-X, Tween-20, and Tween-80 (**Tables 5–7**).

3.3 CO_2 capture by *Oscillatoria* in photo-bioreactor

Photo-bioreactors act as closed pond system that are used for microalgae cultivation, as they can reduce contamination risk from unwanted algae, mold, and bacteria; control temperature; minimize water evaporation; and reduce carbon

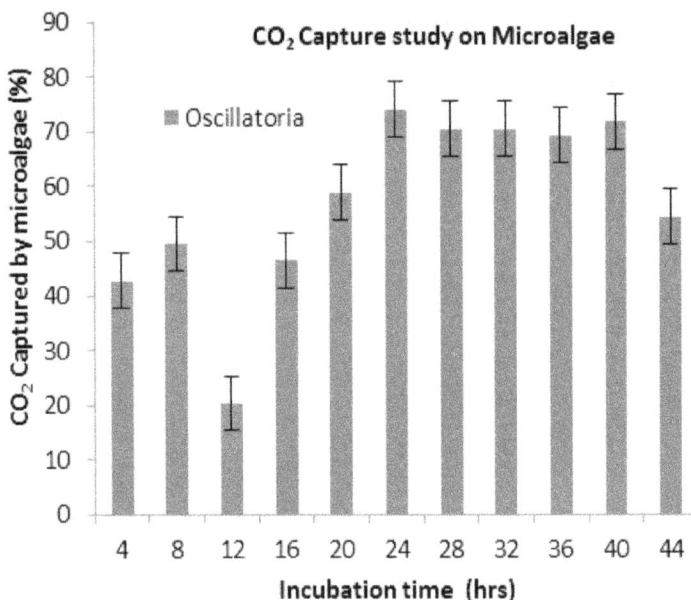

Figure 6.
CO₂ capture by Oscillatoria.

dioxide losses. Photo-bioreactor FMT 150 used in the present study consists of an algal-cultivation vessel with a sealable lid and a base box containing electronics circuitry, LED light panel, and other components essential for the optimal operation of the photo-bioreactor (**Figure 3**). The cultivation vessel is rectangular in shape and flat with a working volume capacity of 400 ml. Its front and back windows are made of glass plates. The bottom of the cultivation vessel is made of stainless steel and contains a thermal bridge that helps heat transfer between a Peltier cell in the instrument base and the culture suspension.

The culture vessel is fixed with the array of high-power light emitting diodes (LEDs). These LEDs produce a highly uniform light with the irradiance flux that can be controlled in the range of 0–3000 µmol (photons)/m⁻/s¹ PAR. The irradiance of the LEDs can be dynamically modulated by the instrument control unit through external computer with software. The photo-bioreactor is equipped with semiconductor light sensor for measuring fluorescence emission and suspension optical density by attenuation of light that was emitting from the LEDs. The solenoid valves are used to switch off the gas supply to the culture (bubbling) during optical measurements. For the supply of fresh medium or buffer, the Peristaltic pumps are used in turbidostat and chemostat mode.

Higher CO_2 capture capacity observed in *Oscillatoria* from 16 to 32 h (**Figure 6**). Higher CO_2 capture efficiency, algal biomass productivity was observed in the selected strains of algae at optimum pH of 7–9 and temperature 25–30°C. The results indicate that the selected blue-green algae *Oscillatoria* can be used for CO_2 capture and biomass production, freshwater algae are ideal candidates for CO_2 capture, when treated with CO_2-rich gas.

4. Conclusion

Developed process for the CO_2 capture will lead to environment-friendly alternatives to chemical CO_2 capture process. The study exclusively relates to integrated methods and systems for utilizing 99.9% CO_2 as a nutrient carbon source along with modified growth media derived from coal mine water for filamentous cyanobacteria culture. The modified algal media was enriched with the addition of Fe, Mg, vitamins, and surfactants for higher photosynthetic efficiency of algae.

Acknowledgements

The authors would like to thank the GAIL (India) Ltd., and CSIR, Govt. of India, for its financial support (SSP 7219 and CSC 102). All the authors express their sincere thanks to Dr. P.K. Singh, Director, CSIR-CIMFR, for supporting the publication of this chapter.

Author details

Vetrivel Anguselvi*, Reginald Ebhin Masto, Ashis Mukherjee
and Pradeep Kumar Singh
Industrial Biotechnology and Waste Utilization, CSIR-Central Institute of Mining
and Fuel Research, Dhanbad, India

*Address all correspondence to: vaselvi@yahoo.com

IntechOpen

References

[1] Kassim MA, Meng TK. Carbon dioxide (CO_2) biofixation by microalgae and its potential for biorefinery and biofuel production. Science of the Total Environment. 2017;**584-585**:1121-1129. DOI: 10.1016/j.scitotenv.2017.01.172

[2] Benemann JR, Van Olst JC, Massingill MJ, Weissman JC, Brune DE. The controlled eutrophication process: Using microalgae for CO_2 utilization and agricultural fertilizer recycling. In: Proceedings of Conference on Greenhouse Gas Technologies; Kyoto, Japan; 2003

[3] Mituya A, Nyunoya T, Tamiya H. In: Burlew JS, editor. Algal cultures from laboratory to pilot plant. Washington: Carnegie Institute; 1953. p. 266

[4] Becker WE, Venkataraman LV. Algae for feed and food. In: A Manual on the Cultivation and Processing of Algae as a Source of Single Cell Protein. Mysore: Wesley Press; 1978

[5] Geitler L, Cyanophyceae L. 14. Band. In: Rabenhorst L, editor. Kryptogamen-Flora. Leipzig: Akademische Verlagsgesellschaft; 1932. p. 1196